37 Advances in Polymer Science

Fortschritte der Hochpolymeren-Forschung

Edited by H.-J. CANTOW, Freiburg i. Br. · G. DALL'ASTA, Colleferro
K. DUŠEK, Prague · J. D. FERRY, Madison · H. FUJITA, Osaka
M. GORDON, Colchester · J. P. KENNEDY, Akron · W. KERN, Mainz
S. OKAMURA, Kyoto · C. G. OVERBERGER, Ann Arbor
T. SAEGUSA, Kyoto · G. V. SCHULZ, Mainz · W. P. SLICHTER, Murray Hill
J. K. STILLE, Fort Collins

With 19 Figures

Springer-Verlag
Berlin Heidelberg GmbH 1980

Editors

Prof. HANS-JOACHIM CANTOW, Institut für Makromolekulare Chemie der Universität, Stefan-Meier-Str. 31, 7800 Freiburg i. Br., BRD

Prof. GINO DALL'ASTA, SNIA VISCOSA – Centro Studi Chimico, Colleferro (Roma), Italia

Prof. KAREL DUŠEK, Institute of Macromolecular Chemistry, Czechoslovak Academy of Sciences, 162 06 Prague 616, ČSSR

Prof. JOHN D. FERRY, Department of Chemistry, The University of Wisconsin, Madison, Wisconsin 53706, U.S.A.

Prof. HIROSHI FUJITA, Department of Polymer Science, Osaka University, Toyonaka, Osaka, Japan

Prof. MANFRED GORDON, Department of Chemistry, University of Essex, Wivenhoe Park, Colchester C04 3 SQ, England

Prof. JOSEPH P. KENNEDY, Institute of Polymer Science, University of Akron, Akron, Ohio 44325, U.S.A.

Prof. WERNER KERN, Institut für Organische Chemie der Universität, 6500 Mainz, BRD

Prof. SEIZO OKAMURA, No. 24, Minami-Goshomachi, Okazaki, Sakyo-Ku, Kyoto 606, Japan

Prof. CHARLES G. OVERBERGER, Macromolecular Research Center, Institute of Science and Technology, The University of Michigan, Ann Arbor, Michigan 48 104, U.S.A.

Prof. TAKEO SAEGUSA, Department of Synthetic Chemistry, Faculty of Engineering, Kyoto University, Kyoto, Japan

Prof. GÜNTER VICTOR SCHULZ, Institut für Physikalische Chemie der Universität, 6500 Mainz, BRD

Dr. WILLIAM P. SLICHTER, Chemical Physics Research Department, Bell Telephone Laboratories, Murray Hill, New Jersey 07 971, U.S.A.

Prof. JOHN K. STILLE, Department of Chemistry, Colorado State University, Fort Collins, Colorado 805 23, U.S.A.

ISBN 978-3-662-15805-0   ISBN 978-3-540-38277-5 (eBook)
DOI 10.1007/978-3-540-38277-5

Library of Congress Catalog Card Number 61-642

© by Springer-Verlag Berlin Heidelberg 1980
Originally published by Springer-Verlag Berlin Heidelberg New York in 1980
Softcover reprint of the hardcover 1st edition 1980

2152/3140 – 543210

Cationic Ring-Opening Polymerization of Heterocyclic Monomers
I. Mechanisms

Stanisław Penczek, Przemysław Kubisa and Krzysztof Matyjaszewski

Centre of Molecular and Macromolecular Studies, Polish Academy of Sciences, Boczna 5, PL-90-362 Łódź

Table of Contents

IV

1 Introduction

The detailed understanding of the chemistry of the elementary reactions of cationic polymerization, including the corresponding kinetic parameters, is only available for the ring-opening polymerization of heterocyclics.

Studies of cationic vinyl polymerizations, started many years ago[1, 2], have not yet yielded the conditions for the quantitative treatment of the elementary reactions involved. This is because the originally formed macrocations not only propagate but also react in many ways, changing their structure and concentration. The large majority of these side reactions are still obscure, even in the polymerization of the best studied monomers, namely isobutene and styrene. Only a few side reactions have been established, for instance the nucleophilic attack of the aromatic ring on the polystyryl carbocation and the formation (with an anchimeric assistance) of the stable 3-phenylindane type end group[3] as well as proton transfer to monomer involving formation of dead macromolecules with unsaturated end groups (isobutene[4], vinyl ethers[5]). These reactions are comprehensively documented and are inherent features of these systems, excluding the possibility of creating living chains with either these or closely related monomers.

Only recently, studies of the non-stationary polymerization of styrene and its derivatives using the flow and stopped-flow spectroscopy have allowed the determination of the rate constants in a more reliable way[6, 7].

When, however, a fast initiation is followed by propagation being the only subsequent reaction (transfer or termination excluded), the quantitative studies of the propagation itself become markedly facilitated[8]. The extensive use of such living kinetic chains provided detailed knowledge of the anionic polymerization of vinyl monomers[8, 9, 10] and, more recently, of the anionic polymerizations of the first two cyclic monomers, namely ethylene oxide and propylene sulfide[11, 12]. Similar conditions leading to living chains were found in the cationic polymerization of heterocyclic monomers polymerizing with ions as reactive species. Onium ions are much more stable than carbocations and thus much more discriminating in their reactions. Therefore, reactions demanding higher activation energies (e.g. transfer) can be eliminated.

The much higher stability of onium ions, compared to that of the majority of carbocations, reflects the fact that, formally, onium ions can be considered to be equivalent to carbocations bonded to (and therefore stabilized by) nucleophilic ligands such as amines, ethers or sulfides. For instance, trimethyloxonium ion $1b$ can be treated as a methylium cation CH_3^+ $1a$ attached to methyl ether and using one of the lone electron pairs of the oxygen atom to fill up the empty orbital of the carbenium ion:

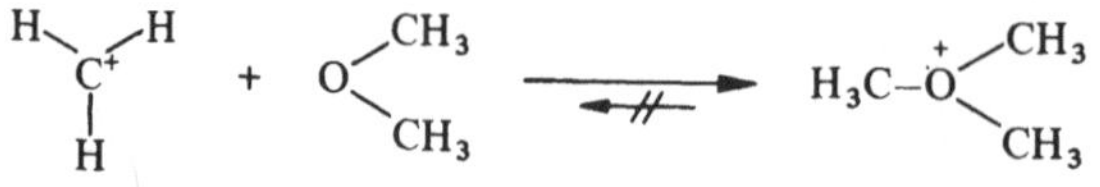

$$\underset{1a}{\underset{\displaystyle H}{\overset{\displaystyle H}{\underset{|}{\overset{\diagup}{C^{+}}\diagdown}}}\overset{\displaystyle H}{}} \quad + \quad \underset{\displaystyle CH_3}{\overset{\displaystyle CH_3}{O\diagdown}} \quad \xrightarrow{\quad\;\not\!/\;\quad} \quad \underset{1b}{H_3C-\overset{+}{\underset{\displaystyle CH_3}{O}}\overset{\displaystyle CH_3}{\diagup}} \tag{1}$$

Some of these ligands are so highly nucleophilic, i.e. basic (see Section II.2), that in an equilibrium like Eq. (1) the carbocation counterpart of the onium cation is not regenerated. Thus, the latter is the only species in solution.

The high stability of the onium ions makes them immune to some reactions that cannot be avoided in the case of carbenium ions. Some onium ions, like tetraalkyl ammonium cations, do not even react with water, which is too weak a nucleophile and cannot displace the amine ligand. This stability, in turn, has allowed the discovery of systems (monomers, initiators, solvents, temperature range) which behave as living systems,

Some cyclic amines, iminoethers, sulfides, ethers, esters, and acetals were indeed demonstrated to polymerize without appreciable transfer and/or termination. It cannot be excluded that for monomers belonging to other compound classes, e.g. lactams or cyclic esters of phosphoric acid, the conditions for the transfer- and termination-less processes will be found in the near future.

In this paper we analyze the cationic polymerizations of cyclic monomers using mostly results of living systems like tetrahydrofuran (THF) which is at present the only comprehensively studied monomer[13]. Other systems are also reviewed and reasons for transfer and termination processes accompanying propagation are analyzed.

2 Monomer Structures, Ring Strains and Nucleophilicities (Basicities)

In this section, which is not intended to be comprehensive, only the most important facts, related to the polymerization processes, are examined. Thus, we include in this part only cyclic ethers, sulfides and amines because these types of monomers have most widely been studied.

2.1 Conformation of Heterocyclic Monomers

The geometry of heterocyclic monomers is governed mainly by the number of atoms in the ring. There are two sources of strain in the rings; one of these is caused by a difference between the angles resulting from a normal orbital overlap and the angles resulting from the number of atoms in the ring. The other arises from the interactions of non-bonded atoms which happen to be located, due to the geometry of the molecule, in close proximity. Distinctions of the bond angles and non-bonding interactions, both causing the ring strain, oblige the heterocycles to assume a conformation for which there is an energy minimum. Due to these distortions, the majority of cyclic molecules, with exception of three-membered ones, are not planar. Thus, even cyclo-butane exists in a twisted conformation with an angle of 20° between the planes. This conformation is mainly caused by non-bonding interactions between hydrogen atoms in the eclipsed conformation.

These interactions are decreased by replacing one CH_2 group by oxygen and oxetane is believed to be planar. Thus, the strain, connected with the abnormally low bond angles within the ring, is important for 3- and 4-membered rings. For 5-membered rings (cyclopentane and its derivatives), the bond angle would be 108° if these rings were flat, and this is very close to the normal value of 109° 28'.

In these and larger rings, the other kinds of strain become dominant. For instance, in THF a considerable repulsion may exist between the non-bonded electron pairs on the oxygen and the electrons engaged in bonding the hydrogen atoms to the nearby carbon atoms. This is a result of the sp^3 hybridization of the oxygen atom supported by the nearly tetrahedral angles. Thus, the free electron pairs behave sterically like a medium sized group. In THF the unshared electrons are constrained to orbitals that are eclipsed by the nearby bonds to hydrogen whereas in THP the orbitals and the

Oxetane   THF   THP

nearby bonds are staggered causing less or no strain. On the other hand, oxepane (hexamethylene oxide) has lost the perfect staggering of the six-membered ring (this leads to its high basicity; see Sect. 2.2).

Still larger rings display a tendency to assume a conformation in which each bond is in gauche conformation in order to minimize the non-bonded interactions present in the eclipsed form. This eventually leads to the transanular strain. For any ring, however, except for 6-membered ones, there is no possible conformation in which both types of strain are relieved. For larger rings ($n > 10$), the above discussed sources of strain may be relieved but then the bond angles within the ring become larger than normal; this introduces an additional strain.

Thus, a molecule of a heterocycle assumes a conformation, in which the sum of all possible interactions, which are a source of strain, is minimized.

In Table 1 the ring strains of some heterocycles and their cycloalkane analogs are listed. The values of the ring strain were calculated as the difference between the calculated and measured enthalpies of formation of the corresponding cyclic compounds.

The monomers most comprehensively studied in cationic polymerization are the non-planar tetrahydrofuran and 1,3-dioxolane. However, 5-membered rings do not assume such favorable conformations like the chair conformation of 6-membered rings. Consequently, the energetic barrier (pseudorotation barrier) between the con-

Table 1. Ring strain (in kcal · mole^{-1}) in heterocyclic monomers[14]

n	$\left(\begin{matrix} CH_2 \\ (CH_2)_{n-1} \end{matrix}\right)$	$\left(\begin{matrix} O \\ (CH_2)_{n-1} \end{matrix}\right)$	$\left(\begin{matrix} CH_2 \\ O \diagdown \diagup O \\ (CH_2)_{n-3} \end{matrix}\right)$	$\left(\begin{matrix} NH \\ (CH_2)_{n-1} \end{matrix}\right)$	$\left(\begin{matrix} S \\ (CH_2)_{n-1} \end{matrix}\right)$
3	27.4	27.3	–	26.9	19.8
4	26.0	25.5	–	–	19.7
5	6.0	5.6	6.2[a]	5.8	1.97
6	– 0.02	1.2	0[a][c]	– 0.15	– 0.3
7	5.1[a]	8.0[b]	4.7[a]	–	3.5[a]
8	8.2[a]	10.0	12.8[a]	–	–

[a]   Heat of polymerization $-\Delta H_{gg}$[15]
[b]   From Ref. 16
[c]   1,3,5-trioxane : $-\Delta H_{LC} = 5.4$ kcal · mole^{-1} [58]

formers with nearly equal energies is low. Among a number of possible conformations, usually two structures are distinguished: the envelope and half chair.

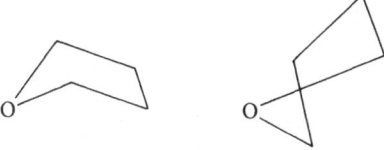

The deviation from planarity is characterized by torsional angles. This angle for tetrahydrofuran is 21° whereas for tetrahydrothiophene a value of 42° is reported[17]. For the 5-membered cyclic acetal, 2,2-dimethyl-1,3-dioxolane, the value of the torsional angle is 20–25°, similar to that in tetrahydrofuran[18].

2.2 Factors Affecting Reactivities

There are three factors that will be separately discussed:
— the nature of the heteroatom, its electronegativity and bond strength with the carbon atom
— the size of the ring
— the steric factors.

In this review we shall use the expressions nucleophilicity and basicity. Nucleophilicity reflects the ability of the species to combine with an electrophilic substrate; it is determined by the kinetically controlled conditions whereas basicity is measured from thermodynamically controlled conditions (from the studies of equilibria). The basicities of a large number of organic compounds have been measured and sometimes we shall use the known order of basicities either to predict or compare the behaviour which should rather be described by an order of nucleophilicities. These are, however, much more specific for a given chemical reaction and no general order of nucleophilicities is known.

The ability of a given compound to form hydrogen bonds is expressed by its basicity and the extent of this ability is usually based on the proportion of the hydrogen bonded compound measured at equilibrium.

These measurements are based on a number of different methods, including the observation of the OD shift ($\Delta\nu$) in the IR region of CH_3OD in its mixture with the studied base, the shift in the UV region of the complexed phenol, the heat of mixing of the studied base with $CHCl_3$, $B(CH_3)_3$ and other non-protonic acids[19].

Basicities are quantitatively expressed by pK_a values. The larger the value of pK_a, the larger the basicity. Some of the pK_a values are given together with $\Delta\nu$ (in cm^{-1}) for CH_3OD in the following order[19]:
n-$C_4H_9NH_2$: 10.8 (260); $C_6H_5NH_2$: 4.8 (180); N,N-dimethylacetamide: 0.0 (120); THF: −2.1 (117); $(C_2H_5)_2O$: −3.6 (96); $C_6H_5OCH_3$: −6.5 (70); $(CH_3)_2CO$: −7.1 (75); $C_6H_5NO_2$: −11.5 (25).

Hydrogen bonding with CH_3OH and C_6H_5OH used as proton donors, gives the following order of basicity: $R_3N > R_3P > R_2O > R_2S$. However, R_2S becomes more "basic" than R_2O when the ability of bonding with softer acids is measured (e.g. $SbCl_5$, $SnCl_4$ of I_2). Alcohols are usually more basic than the corresponding ethers ($CH_3OH > (CH_3)_2O$), and the order or basicities of amines, according to substitution, is as follows: NH_3 > primary > secondary > tertiary although the differences are rather small and the order given above is not very certain (pK_a: NH_3 (9.2), $(CH_3)_3N$ (9.8)).

The order of basicities of ethers is the following:

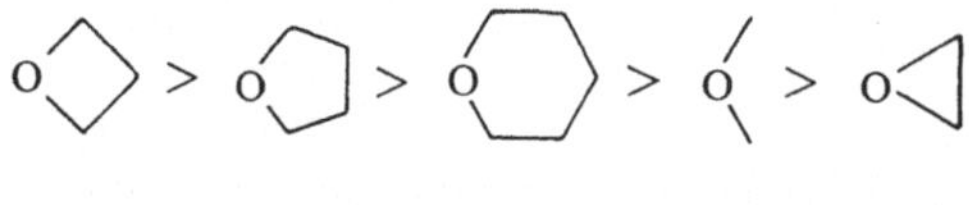

whereas for cyclic sulfides the open-chain compounds were found to be more basic than the cyclic ones. The OD shifts of CH_3OD (in cm^{-1}) and heats of mixing with $CHCl_3$ in 1:1 solutions (in $kcal \cdot mole^{-1}$) at 25 °C for several cyclic ethers are given below[19].

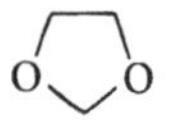

76; –

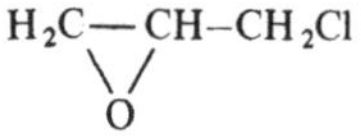

80; 0.19

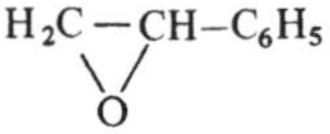

85; 0.24

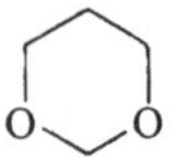

85; –

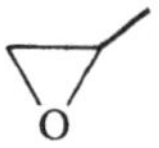

99; 0.47

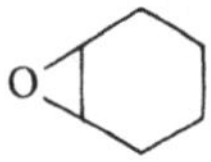

99; 0.67

$(C_2H_5)_2O$

96; 0.70 ($pK_a = -3.6$)

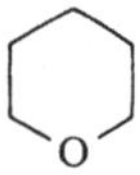

115; 0.60

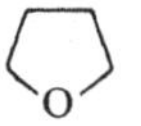

117; 0.68; ($pK_a = -2.1$)

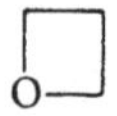

120; 0.70

Among factors affecting the basicities of the studied compounds the following, which we discuss on the basis of Searles' review, are the most important[20]:

Inductive Effects: increasing the positive inductive effect tends to increase the donor ability (cf. $(C_2H_5)_2O$ and $(CH_3)_2O$);

Conjugation, due to the interaction of a lone pair of electrons on the heteroatom with an adjacent π-orbital, diminishes the ability of the heteroatom to coordinate with an acceptor molecule.

Steric effects: the presence of bulky groups close to the heteroatom or on the acceptor molecule may inhibit the interaction.

The ring size affects the basicity in various ways: a decrease in the angle, involving a heteroatom associated with a decrease in the ring size, results in less steric hindrance to interaction; changes in its valence angle result in a variation of the electron density on the heteroatom.

The fact that e.g. tetrahydrofuran is a better electron donor than acyclic ethers is in accord with its lower steric requirements and stems from the effect of the bond angles between oxygen and adjacent carbon atoms on the electron density of the oxygen atom.

These differences due to the steric requirements and lone pair repulsion by σ electrons of the adjacent bonds may disappear for larger heteroatoms and other effects may prevail. This may be among the reasons for the higher basicity of linear as opposed to cyclic sulfides.

The values of basicities discussed above, determined from thermodynamic measurements, are sometimes used to estimate the nucleophilicities revealed in kinetic studies. Although this is a rather crude approach it has yielded a number of satisfactory results explaining various phenomena of the cationic polymerization of heterocycles (e. g. the order of the overall reactivities in cationic copolymerization).

3 Initiation

Initiation of the cationic polymerization of heterocycles can be accomplished in many ways. This is in contrast to the propagation step which always consists of reactions of macrocations with monomer molecules. Thus, by improving the initiators used, i.e. if new initiators giving fast, quantitative initiation and producing stable growing species are found, some of those employed earlier will become interesting only historically. In our opinion, many initiators used for the cationic polymerization of heterocycles have at present merely this weight. Nevertheless, all of the major classes of initiators are covered in this review and only in the last part of this section, summarizing the chemistry of initiation, initiators are selected that can advantageously be used either in screening the polymerization of new monomers, when high molecular products are desired, or in the studies of propagation mechanisms.

3.1 Protonic Acids

Protonic acids could appear to be the best candidates for simple, clean initiation. Reactions frequently written as e.g.:

$$\text{HA} + \text{O}\bigcirc \longrightarrow \text{H}-\overset{+}{\text{O}}\bigcirc \ \text{A}^- \tag{2}$$

are however much more complicated and involve a number of equilibria.

Protonic acids, dissolved in polar solvents, can form the corresponding dimers, which can dissociate etc.:

$$\text{HA} \ \overset{\text{HA}}{\rightleftharpoons} \ (\text{HA})_2 \ \rightleftharpoons \ \text{H}_2\text{A}^+\text{A}^- \ \overset{(\text{HA})_2}{\rightleftharpoons} \ \text{H}_2\text{A}^+(\text{HA})_2 + \text{A}^- \tag{3}$$

Only recently, the application of the polarographic method by Plesch provided quantitative information on some of the equilibria given in Scheme (3)[21]. It was found, for instance, that the equilibrium constant K of binary ionogenic reactions:

$$2\,\text{HA} \overset{K}{\rightleftharpoons} \text{H}_2\text{A}^+ + \text{A}^-$$ equals $K = 7.2 \cdot 10^{-3}$ (CF$_3$SO$_3$H) and $K = 9.8 \cdot 10^{-2}$ (FSO$_3$H) in CH$_2$Cl$_2$ at $23 \pm 2\,°$C.

These reactions, which are not well known in organic solvents, are important for the polymerization of monomers of low nucleophilicity (e.g. styrene[22], 1,1-diphenylethylene[23]) but can apparently be neglected when a strong nucleophile is present. For this group of monomers, the following equilibria must be considered:

$$HA + X \bigcirc \rightleftharpoons AH \cdots X \bigcirc \rightleftharpoons A^-, H - \overset{+}{X} \bigcirc \rightleftharpoons A^- + H - \overset{+}{X} \bigcirc$$

$$(4)$$

In Eq. (4) only one monomer molecule is shown in form of the H-bonded complex or the secondary oxonium ion. It is known, however, that when a strong protonic acid is dissolved in water the hydronium ion H_3O^+ is formed which is specifically solvated by three molecules of water[24]:

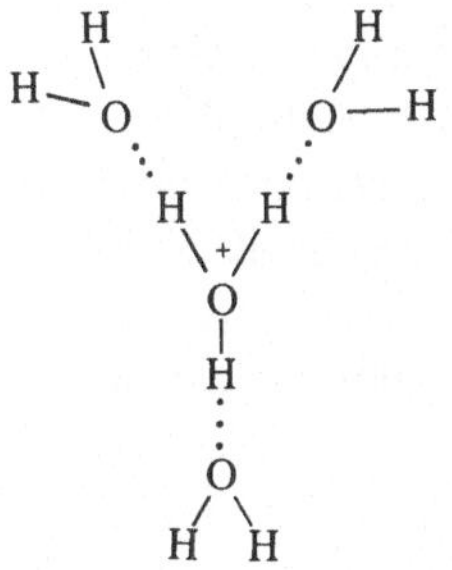

$$(5)$$

It has also been shown, that the interaction of one molecule of acid (HBr) with one molecule of a weak organic base (e.g. THF) in chlorinated hydrocarbons leads predominantly to the formation of a hydrogen bonded complex. A second molecule of the base is needed in order to convert this complex into the secondary oxonium ion. However, the equilibrium constant of the ionization reaction is still low[25]:

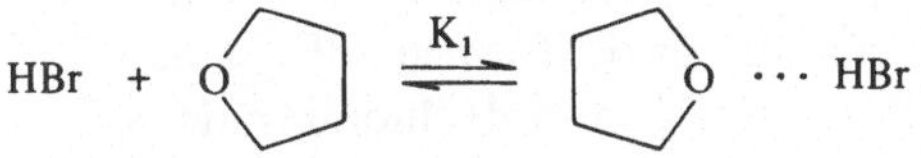

$$HBr + \bigcirc O \overset{K_1}{\rightleftharpoons} \bigcirc O \cdots HBr \qquad K_1 = 48.5 \; mole^{-1} \cdot l \qquad (6)$$

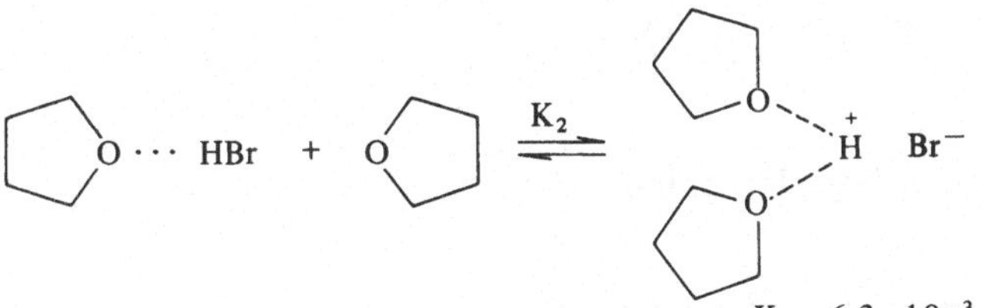

$$\bigcirc O \cdots HBr + \bigcirc O \overset{K_2}{\rightleftharpoons} \qquad (7)$$

$$K_2 = 6.3 \cdot 10^{-3} \; mole^{-1} \cdot l$$
$$(CHCl_3, 25°C)$$

Of course, the ratio of the proton affinities of the monomer used to those of the anion determines the nature and position of the equilibrium.

The data given above are the only quantitative data available for the interaction of strong protonic acids with organic bases in non-aqueous solutions. There is, however, immense literature available on the interaction of bases with weak acids. Quantitative measurements of these interactions were used for the estimation of the basicity which is a measure of the ability to form hydrogen bonds with hetero-atoms[19].

3.1.1 Protonic Acids with Complex Anions

Combination of e.g. HF with BF_3 or SbF_5 in a polar solvent should formally result in the formation of the corresponding protonic acid:

$$HX + MtX \rightleftharpoons HX \cdot MtX_n \rightleftharpoons H^+(MtX_{n+1})^- \tag{8}$$

Mt = metal

These processes are complicated by the fact that acids with complex anions exclusively exist in the ionized form and, therefore, ionization of the complex $HX \cdot MtX_n$ requires stabilization of the proton by solvation which becomes possible in the presence of the corresponding nucleophile. In this way, a number of secondary oxonium salts with the complex $SbCl_6^-$ anion were prepared (and isolated as crystalline compounds) by mixing HCl with $SbCl_5$ in a proper solvent in the presence of an oxygen base[26].

The stability of the complex anion depends on the nature of both Mt and X. The stability of complex anions decreases in the following order[27, 28]:

$$SbF_6^- \sim AsF_6^- > PF_6^- > SbCl_6^- > BF_4^- > AlCl_4^- \tag{9}$$

In general, this stability depends on geometric (crowding of halogen atoms) and electronic features (electronic structure of the central atom and its ability to acquire the negative charge).

Thus, even relatively weak bases are protonated by the mixture $HF - SbF_5$ because of the low nucleophilicity and high stability of the formed SbF_6^- anion[29]. The stability is mostly due to the high affinity of SbF_5 towards fluoride anions, F^-. On the other hand, the proportion of protonation is much lower when less stable and more nucleophilic anions are used (e.g. $AlCl_4^-$).

3.1.2 Protonic Acids with Non-Complex Anions

If in Eq. (4) A^- denotes a non-complex anion, e.g. Cl^-, CF_3COO^-, ClO_4^- or $CF_3SO_3^-$, then the competition between the recombination of counterions and propagation has to be considered because these anions may recombine with growing macrocations to give covalent bonds. This is the major difference between non-complex anions and complex anions, the latter being unable to form such chemical bonds. Thus, in the polymerization with non-complex anions, recombination after a certain number of propagation steps (e.g. n steps in Scheme (10)) can be expected:

$$H - \overset{+}{X} \bigcirc \quad A^- + n \; X \bigcirc \quad \underset{\rightleftharpoons}{\overset{k_p}{\longrightarrow}} \quad H - (X\text{---})_n \; \overset{+}{X} \bigcirc \quad A^- \quad \overset{k_t}{\longrightarrow} \quad H - (X\text{---})_{n+1} A \tag{10}$$

(k_p can acquire discrete values in the first addition steps)

The degree of polymerization n of the polymer formed would thus depend on the ratio of the rate constants of propagation (k_p) and recombination of the macro-cation with the anion (the counterion) (k_t). This ratio k_p/k_t is, besides other less important factors, proportional to the ratio of the nucleophilicity of the monomer to that of the anion. Thus, a given anion may be unable to sustain the polymerization of a monomer of low nucleophilicity whereas the same anion can recombine only after thousands of propagation steps of the more nucleophilic monomer. The chloride anion can be used as an example: this anion cannot give rise to the polymerization of cyclic acetals or ethers because simple addition of the initiator (e.g.HCl) to the first monomer molecule would take place. The formation of $ClCH_2CH_2OH$ from HCl and ethylene oxide (n = 1 in Eq. (10)) is an example[30]. On the other hand, highly nucleophilic N-substituted cyclic amines can be successfully polymerized to high polymerization degrees with Cl^- anions.

In Eq. (10) recombination of counterions and formation of the covalent macro-molecule is described as an irreversible reaction. However, this can be a reversible process in which ionization involves nucleophilic attack of a monomer molecule or a polymer segment.

One of the few systems for which the initiation with protonic acid has been comprehensively studied is the cationic polymerization of lactams[31]. Initiation of the cationic polymerization of lactams with HCl gives predominantly a monomer molecule protonated at the oxygen atom but a small amount of N-protonated lactam is assumed to be present in the tautomeric equilibrium[32]:

$$\left[\underset{11a}{HO\underset{\bigcirc}{\overset{+}{C}-NH}} \longleftrightarrow HO\underset{\bigcirc}{\overset{+}{C}=NH} \longleftrightarrow HO^+\underset{\bigcirc}{\overset{\Vert}{C}-NH} \right] \rightleftharpoons \underset{11b}{O\underset{\bigcirc}{\overset{\Vert}{C}-\overset{+}{N}H_2}} \tag{11}$$

For the sake of simplicity the protonated monomer molecule is further represented by *11b*. (See however Addendum p. 131). The acylation of the monomeric lactam results in the formation of an aminoacyllactam:

$$\underset{\bigcirc}{\overset{+}{CO-NH}} + \underset{\bigcirc}{\overset{+}{CO-NH_2}} \rightleftharpoons \underset{\bigcirc}{CO-N-CO} \; \overset{+}{NH_3} \tag{12}$$

The protonated primary amino group is in equilibrium with the protonated lactam:

$$\cdots - \; \overset{+}{NH_3} + \underset{\bigcirc}{CO-NH} \rightleftharpoons \cdots - \; NH_2 + \underset{\bigcirc}{\overset{+}{CO-NH_2}} \tag{13}$$

but the positions of the equilibria postulated above have not been measured.

The reactions described above were deduced from the structures of the oligo-meric products separated by electrophoresis and identified by comparison with inde-pendently synthesized compounds.

3.1.3 Direct Identification of the Initiation Products (Initial Species)

Polymerization of THF initiated with fluorosulfonic acid (FSO$_3$H) was studied by ^{1}H-NMR in C$_6$H$_6$[33]. A singlet due to the acidic proton, exchangeable during poly-merization between the undissociated H-bonded acid, secondary oxonium ions and the polymer bearing hydroxy end groups, shifts with the monomer conversion from about 16 ppm to about 11 ppm, the intensity remaining invariable (interchangeable protons are underlined):

$$\underline{\text{FSO}_3\text{H}} + \text{O}\bigcirc \rightleftharpoons \underline{\text{FSO}_3\text{H}}\cdots\text{O}\bigcirc \rightleftharpoons \text{FSO}_3^-, \text{H} - \overset{+}{\underline{\text{O}}}\bigcirc$$

$$\underset{14a}{} \qquad\qquad \underset{14b}{} \qquad\qquad \underset{14c}{} \qquad\qquad (14)$$

$$\underline{\text{H}}\overset{+}{\underset{}{\text{O}}}\bigcirc \overset{n\,\text{THF}}{\rightleftharpoons} \underline{\text{H}}\!-\!\!\big[\text{O}\!-\!(\text{CH}_2)_4\big]_{\overline{n}}\!\!-\!\overset{+}{\text{O}}\bigcirc$$

Oxygen atoms in macromolecules can also be protonated by any species $14a-14c$; e.g.:

$$\text{FSO}_3^-, \text{H}-\overset{+}{\text{O}}\bigcirc + \begin{array}{l}\text{CH}_2\!-\!\text{CH}_2\!-\cdots\\|\\\text{O}\!-\!\text{CH}_2\!-\cdots\end{array} \rightleftharpoons \text{FSO}_3^-, \text{H}-\begin{array}{l}\text{CH}_2\!-\!\text{CH}_2\!-\cdots\\|\\\overset{+}{\text{O}}\!-\!\text{CH}_2\!-\cdots\end{array} + \text{O}\bigcirc \quad (15)$$

Only 1:1 H-bonded complexes are shown in Schemes (14) and (15) although the actual state of solvation in systems with a large excess of monomer can involve more than one molecule (cf. Eqs. (6) and (7)).

The averaged chemical shift of exchangeable protons moves upfield with conversion and should eventually stabilize after the polymer − monomer equilibrium is attained and when $\overline{\text{DP}}_n$ becomes constant. This is because initiation with the secondary oxo-nium ions is slow and first of all, a polymer of higher $\overline{\text{DP}}_n$ is formed. Polymer − mono-mer equilibrium is reached for these longer chains which, through depropagation, will shorten, the number of hydroxy end groups increasing until the final distribution (constant $\overline{\text{M}}_w/\overline{\text{M}}_n$) is reached.

Recently, the initiation of 1,3-dioxolane with CF$_3$SO$_3$H in CH$_2$Cl$_2$ was studied by ^{1}H-NMR[34]. The solution of CF$_3$SO$_3$H in CH$_2$Cl$_2$ shows an absorption for CF$_3$SO$_3$$H$ at 8.3 ppm at a concentration of the acid equal to 10^{-1} mole $\cdot$ l^{-1}. The state of the acid in CH$_2$Cl$_2$ is unknown (cf. also Ref. 21), but on the basis of this chemical shift one can assume that there is a large proportion of the covalently bonded acid. Upon addition of 1,3-dioxolane to this solution the chemical shift of acidic proton moves downfield and stabilizes at 14.5 ppm when the ratio of the components [1,3-dioxolane]/[CF$_3$SO$_3$H] is close to unity. A further increase of this ratio does not appreciably change the chemical shift until [1,3-dioxolane]$_0$ reaches its equilibrium concentration and the monomer starts to polymerize. Apparently, in this system the following equilibria operate before the onset of polymerization:

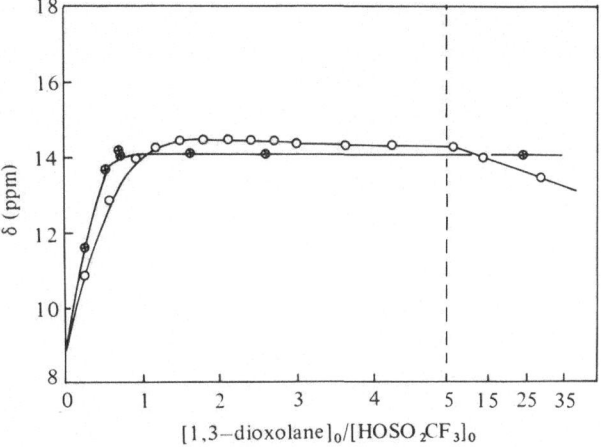

$$CF_3SO_3H \; + \; \overset{\frown}{O \underset{\smile}{\quad} O} \; \underset{K_H}{\rightleftharpoons} \; CF_3SO_3H \ldots \overset{\frown}{O \underset{\smile}{\quad} O} \; \underset{K_i}{\rightleftharpoons} \; CF_3SO_3^- \ldots H - \overset{+}{O} \underset{\smile}{\quad} O$$

(16)

(or $CF_3SO_3^-$ H)

Fig. 1. Dependence of the chemical shifts of acidic proton on the ratio of concentrations [1,3-dioxolane]/[HOSO$_2$CF$_3$] (○) and [(C$_2$H$_5$)$_2$O]/[HOSO$_2$CF$_3$] (⊛) in CH$_2$Cl$_2$ at 35 °C

These changes observed in ^{1}H-NMR are illustrated in Fig. 1. The same figure describes the chemical shift of the acidic proton as a function of the ratio of concentrations [(C$_2$H$_5$)$_2$O]/[HOSO$_2$CF$_3$] for a non-polymerizable ether. This system qualitatively behaves in the same way as a system with 1,3-dioxolane; however, obviously, no upfield shift (due to polymerization and formation of HO—CH$_2$— . . . groups) is observed for diethyl ether.

At 25 °C, the rate of exchange of protons between components involved in Eq. (16) disallows the determination of their proportions although it follows from the dependence of the chemical shift on the starting ratio of 1,3-dioxolane to acid that the first equilibrium is almost completely shifted to the right-hand side.

The individual steps of the equilibria as in Scheme (16) may depend on monomer structure, i.e. the number of heteroatoms in the monomer molecule, their basicity, the strength of H-bonding and the acidity (pK$_a$) of the protonic acid.

There are no data available on the rate of formation of dialkyloxonium ions like protonated 1,3-dioxolane in Eq. (16). It is remarkable that the rate constants of formation of secondary onium ions from linear ethers, acetals, sulfides etc. are also unknown. These should however be lower than the rate constants of proton transfer in water (an upper limit) being close to 10^{11} mole$^{-1}\cdot$l$\cdot$s^{-1} [35], but certainly higher than the rate constants of protonation of olefins with CF$_3$SO$_3$H; reported values are as follows: k $\simeq 10^3$ mole$^{-1}\cdot$l$\cdot$s^{-1}(−80 °C) for 1,1-diphenylethylene[36] and k $\simeq 10^1$ mole$^{-1} \cdot 1 \cdot$ s^{-1} for styrene (30 °C)[37] (the difference looks unexpectedly large).

The major difference between the initiation of heterocyclic and vinyl monomers studied until now is grounded on the efficiency of initiation. It has been conclusively shown that every acid molecule is consumed in the polymerization of 1,3-dioxolane[34] whereas only a portion of the acid is used in the initiation of vinyl monomers, e.g. one CF_3SO_3H molecule for every three applied to 1,1-diphenylethylene[38]. It is possible that in vinyl polymerization, the acid itself has to stabilize the components of the ion pair or the anion in the initiation step (it was, for instance, assumed that two CF_3SO_3H molecules stabilize one $CF_3SO_3^-$ anion)[38]. In the polymerization of much more nucleophilic heterocyclic monomers, this stabilization is provided by monomer. Perhaps, vinyl monomers containing heteroatoms could also provide stabilized ion pairs upon initiation.

The formation of addition products and further ionization according to Eq. (17) were shown directly also in the polymerization of cyclic amines.

^{1}H-, ^{11}B-, and ^{19}F-NMR spectra were utilized by Enikolopyan[39] for the identification of the protonated conidine *17c* obtained from *17a* and a $CH_3OH \cdot BF_3$ complex in CH_3OH solution[39]:

$$\tag{17}$$

The formation of the $BF_3OCH_3^-$ anion has been detected by comparing ^{19}F-NMR spectra of the reaction mixture with that of $K^+BF_3OCH_3^-$ (δ = 156.1 ppm from CFCl$_3$, ^{1}J$_{BF}$ = 12 Hz), whereas formation of *17b* has been established by comparing with a spectrum of a complex *17b* obtained in the absence of the proton donating compounds (δ = 162.5 ppm from CFCl$_3$, ^{1}J$_{BF}$ = 17 Hz).

The protonated form *17c* was identified by its analogy with the hydrobromide of *17a*. Fast exchange of proton between *17c* and CH_3OH at the temperature of measurements did not permit the positions of the equilibria involved in Scheme (17) to be established.

Cyclic ethers, acetals and cyclic amines belong to the group of hard bases while cyclic sulfides can be considered as soft bases. On the other hand proton is the hardest acid whereas alkyl cations are much softer. Thus, in good agreement with the concept of hard and soft acids and bases (the alike ones react faster giving more stable products), secondary oxonium ions react slower with monomer than their tertiary oxonium counterparts. The same is observed for cyclic amines[40] whereas secondary sulfonium ions apparently react faster with cyclic sulfides than the corresponding tertiary sulfonium ions do[41].

3.2 Initiation with Stable Carbenium or Onium Ions and with Their Covalent Precursors ("Hidden" Ions)

The following groups of initiators will be discussed in this section (examples are given in brackets):

1. R_3C^+ $((C_6H_5)_3C^+A^-)$

   carbenium ions

2. $ROCH_2^+$ $(CH_3O\overset{+}{C}H_2A^-)$

   alkoxycarbenium ions

3. $RC\equiv\overset{+}{O}\ A^-$ $(C_6H_5C\equiv\overset{+}{O}\ A^-)$

   oxocarbenium ions

4. $R_n\overset{+}{X}A^-$ $((C_2H_5)_3\overset{+}{O}\ A^-)$

   onium ions (X = N, O, P, S, n = 3 or 4)

5. RA $(CH_3OSO_2CF_3, (CF_3SO_2)_2O)$

   (or even CH_3I for strongly nucleophilic monomers)

   covalent initiators

6. MtX_n (BF_3)

   Friedel-Crafts (F-C) catalysts.

7. Cations formed by electron transfer

8. Organometallic initiators

   $((i\text{-}C_4H_9)_3Al/H_2O$ system)

3.2.1 Carbenium Ions

The best known stable carbocation salts are triphenylmethylium salts, $(C_6H_5)_3C^+A^-$, initiating polymerization of certain olefins by direct addition:

$$(C_6H_5)_3C^+A^- + CH_2=\underset{R}{CH} \rightarrow (C_6H_5)_3C-CH_2-\underset{R}{\overset{+}{C}H}\ A^- \tag{18}$$

The occurrence of the direct addition has been verified in some cases by the determination of the triphenylmethyl moiety in the polymer chain (UV, IR, NMR[43, 44]). Trityl salts can easily be observed in solution because of their specific and strong absorption ($\epsilon_{max} = 3.6 \cdot 10^4$ at λ_{max} = 430 nm for $(C_6H_5)_3C^+A^-$ where A^- is a complex anion).

The corresponding dissociation constants of trityl salts and related thermodynamic parameters are given in Table 2.

In vinyl polymerization, the kinetics of initiation was studied by Sigwalt and Vairon for cyclopentadiene[44] and p-methoxystyrene[49] using the spectrophotometric method.

Stable trityl salts, which might lead to systems devoid of side reactions, do not however initiate polymerization of the majority of heterocyclic monomers by direct addition[50, 51]. The only reported exception is the copolymerization of THF with propylene oxide[52].

Table 2. Dissociation constants and thermodynamic parameters of dissociation of triphenyl-methylium salts in CH_2Cl_2 at 25 °C

Anion	$K_D \cdot 10^4$ (mole $\cdot$ 1^{-1})	ΔH_D (kcal $\cdot$ mole^{-1})	ΔS (cal $\cdot$ mole^{-1} $\cdot$ K^{-1})	Ref.
AsF_6^-	1.4	-2.2 ± 0.2	-25.1 ± 0.8	45)
AsF_6^-	2.2	–	–	46)
SbF_6^-	1.6	-2.2 ± 0.5	-25 ± 2	45)
SbF_6^-	2.5	–	–	46)
$SbCl_6^-$	1.4	-0.8 ± 0.3	-17	47)
$SbCl_6^-$	1.9	-2.0	-23	48)

For THF and 1,3-dioxolane (DXL), it has been shown that in the first step, preceding the true initiation, a hydride ion is abstracted from the monomer molecule[52, 53]. These reactions were known earlier in organic chemistry and will be discussed below.

The following kinetic scheme was proposed to explain the absence of a direct addition (the anions are omitted and 1,3-dioxolane is used as a model monomer)[51].

$$(C_6H_5)_3C^+ \quad + \quad O\!-\!O$$

$$k_H \Big\Vert\, k_{-H} \qquad\qquad k_a \Big\Vert\, k_d \quad K_e = k_a/k_d$$

$$(C_6H_5)_3CH \; + \; O\overset{+}{\cdots}O \qquad (C_6H_5)_3C\!-\!{}^+\!O\!-\!O \quad \xrightarrow[\text{(first propagation step)}]{+\,O\!-\!O \;;\, k_{p1}}$$ (19)

$$O\overset{+}{\cdots}O \;+\; O\!-\!O \;\xrightarrow{k_i}\; \overset{O}{\underset{H}{\overset{\Vert}{C}}}\!-\!O\!-\!CH_2\!-\!CH_2\!-\!{}^+\!O\!-\!O$$

$$\xrightarrow{+\,O\!-\!O} \quad \text{(propagation)}$$

For Scheme (19) K_e, k_H and k_i were measured: $K_e = 6.3 \cdot 10^{-2}$ mole$^{-1} \cdot$ l, $k_H = 1.2 \cdot 10^{-3}$ mole$^{-1} \cdot$ l $\cdot$ s^{-1} (both in CH_2Cl_2 at 25 °C), $k_i = 2 \cdot 10^{-4}$ mole$^{-1} \cdot$ l $\cdot$ s^{-1} (in CD_3NO_2 at 25 °C). Thus, only the small portion of ions existing in the oxonium form is required for further propagation, in spite of the 10^3–10^4 fold excess of monomer over initiator. Moreover, k_{p1} seems to be (it could not be measured) much smaller than k_H. Apparently, the addition of monomer to the oxonium ion bearing the triphenylmethyl moiety is slow for 1,3-dioxolane because of the large steric hindrance preventing the attack of monomer at C-2 in the oxonium ion. Of course, this first step of propagation would be much slower than the subsequent steps, when the triphenylmethyl moiety would be further apart from the site of the active species ($k_{p1} \ll k_p$). Apparently, the use of propylene oxide, yielding less sterically hindered ions could give $k_p > k_H$, as already discussed above.

Recently, the next homolog of the trityl cation, namely the diphenylmethylium cation, was used to initiate polymerization of THF:

$$\underset{C_6H_5}{\overset{C_6H_5}{\diagdown}}\underset{H}{\overset{|}{C^+}}\ SbF_6^- \ + \ n\ O\!\!\bigcirc\ \longrightarrow\ \underset{C_6H_5}{\overset{C_6H_5}{\diagdown}}\underset{H}{\overset{|}{C}}-[O-(CH_2)_4\!\!-]_{\overline{n}}\ \cdots \quad (20)$$

Studies of the polymerization kinetics and UV spectra of the resulting polymers have revealed that initiation by addition proceeds rapidly ($k_i > k_p$) and quantitatively[54]. This result indicates that a decrease of the steric hindrance at the first formed oxonium ion facilitates the second monomer addition which successfully competes with hydride transfer.

Since the actual values of K_e and k_H can indicate whether initiation proceeds through addition or hydride transfer, it is instructive to examine these values.

3.2.2 Carbenium Ion-Onium Ion Equilibria

In spite of many publications in which equilibria of the type described in Scheme (19) are postulated, accepted or rejected on various grounds, the carbenium-onium equilibrium has not been determined quantitatively. Only recently has this equilibrium been studied in a model system, using triphenylmethylium cation as a model carbenium ion and various linear or cyclic ethers or acetals as nucleophiles[55]. At room temperature in CH_2Cl_2, the equilibrium constant was found to be $K_e = 1.6\ mole^{-1}\cdot l$ and $K_e = 6.3\cdot 10^{-2}\ mole^{-1}\cdot l$ for THF and 1,3-dioxolane, respectively. These values imply that, in order to convert 50% of triphenylmethylium cation (at the initial concentration of $10^{-4}\ mole\cdot l^{-1}$) into its oxonium counterpart bearing a THF moiety

$$(C_6H_5)_3C^+ \ + \ \begin{matrix} THF \\ or \\ 1,3-dioxolane \end{matrix}\ \underset{k_d}{\overset{k_a}{\rightleftharpoons}}\ (C_6H_5)_3C-{}^+O\!\!\bigcirc \qquad (21)$$

$$K_e = k_a/k_d$$

it is necessary to use a 10^4 fold excess of THF. A much higher concentration of the less nucleophilic 1,3-dioxolane is needed for the same purpose.

Table 3 contains values of K_e° and of the corresponding thermodynamic parameters ΔH_e° and ΔS_e°.

It has been shown that ΔH_e° is a linear function of basicity expressed by pK_a or pK_b and that ΔH_e° and ΔS_e° are interrelated by the linear isoequilibrium (compensation) plot[55].

3.2.3 Structure of Monomers and Rate of Hydride Transfer

There are only a few reports on the kinetics of hydride transfer. These reactions, studied by Meerwein who established the structure of the products of the hydride

Table 3. Equilibrium constants and thermodynamic parameters of carbenium-oxonium equilibria:

$$(C_6H_5)_3C^+ + O\!\!<\;\overset{K_e}{\rightleftharpoons}\;(C_6H_5)_3C\text{-}\overset{+}{O}\!\!<\quad \text{(Ref. 55)}$$

Substrate	K_e mole$^{-1} \cdot l^a$ 25 °C	-73 °C	ΔH_e^0 (kcal $\cdot$ mole^{-1})	ΔS_e^0 (cal $\cdot$ mole$^{-1} \cdot$ K^{-1})
Water	$6.2 \cdot 10^1$	$1.4 \cdot 10^5$	-9.3 ± 0.3	-23 ± 1
2-Trichloromethyl-1,3-dioxolane	$8.5 \cdot 10^{-2}$	$1.1 \cdot 10^1$	-5.8 ± 0.2	-24 ± 1
1,3-Dioxolane	$3.2 \cdot 10^{-2}$	$1.3 \cdot 10^1$	-7.3 ± 0.6	-31 ± 2
Dimethoxymethane	$2.2 \cdot 10^{-2}$	$2.2 \cdot 10^1$	-8.3 ± 0.2	-35 ± 1
Diethoxymethane	$3.2 \cdot 10^{-1}$	$6.0 \cdot 10^1$	-9.1 ± 0.4	-37 ± 2
Dibutyl ether	$6.7 \cdot 10^{-1}$	$1.1 \cdot 10^3$	-9.0 ± 0.2	-31 ± 1
2-Methyl-1,3-dioxolane	$1.2 \cdot 10^{-2}$	$4.7 \cdot 10^1$	-10.0 ± 0.2	-42 ± 1
2-Phenyl-1,3-dioxolane	$4.6 \cdot 10^{-1}$	$2.2 \cdot 10^3$	-10.1 ± 0.6	-35 ± 2
1,3-Dioxane	$2.0 \cdot 10^{-2}$	$1.8 \cdot 10^2$	-11 ± 1	-44.7 ± 0.6
1,4-Dioxane	$4.6 \cdot 10^{-2}$	$4.5 \cdot 10^2$	-11.1 ± 0.7	-42 ± 5
Tetrahydrofuran	1.6	$1.8 \cdot 10^5$	-14.0 ± 0.8	-46 ± 3
Diethyl ether	$2.3 \cdot 10^{-1}$	$2.9 \cdot 10^5$	-17	-60
Tetrahydropyran	$2.3 \cdot 10^{-1}$	$1.0 \cdot 10^6$	-17 ± 1	-60 ± 5

a   No influence of the ion pairing or anion structure was observed

transfer from dioxolanes to the triphenylmethylium[56] cation, usually lead to a single product (dioxolenium salts) in a high yield[57].

Hydride transfer rate constants were measured only recently by Stomkowski and the first published values of k_H[43, 59] were later confirmed and complemented by other authors[60, 61]. It is interesting to note that k_H determined spectrophotometrically (the rate of disappearance of $(C_6H_5)_3C^+$) agrees well with the results obtained by the polarographic method elaborated by Plesch[60].

In Table 4 some values of k_H for various cyclic ethers and acetals are given and compared with k_H values of their linear counterparts.

It has been shown independently that ion pairing has no influence on the values of k_H[63].

Attempts to measure k_H for the following cyclic acetals and ethers were unsuccessful because of the too low value of k_H (below 10^{-5} mole$^{-1} \cdot l \cdot s^{-1}$ at 25 °C): 2-Trichloromethyl-1,3-dioxolane, 1,3,5-trioxane, tetrahydropyran, and 1,4-dioxane. The non-planar and puckered structure of the last three compounds can hinder a sufficiently close approach to the central atom of the triphenylmethylium cation necessary for hydride transfer to proceed[64].

Correlations between the kinetic parameters and chemical shifts of the hydrogen atoms to be transferred as hydride ions (Table 4) reveal that for the more shielded hydrogen atoms the rate constant decreases. Apparently, this correlation is only valid for compounds having similar steric surroundings of the hydrogen atom to be transferred.

Table 4. Rate constants and activation parameters of hydride transfer from linear and cyclic ethers and acetals to the $(C_6H_5)_3C^+$ cation

Donor	k_H mole^{-1} · l · s^{-1} 25 °C (CH$_2$Cl$_2$)	$\Delta H^{\ddagger}$ (kcal · mole^{-1})	$\Delta S^{\ddagger}$ (cal · mole^{-1} · K^{-1})	^{1}H-NMR δ (in ppm)	Ref.
Tetrahydrofuran	$6.3 \cdot 10^{-3}$	12 ± 1	-26 ± 5	3.6	62)
	$3.9 \cdot 10^{-3}$ (18 °C)	—	—	—	60)
Diethyl ether	$3.0 \cdot 10^{-4}$	10 ± 1	-40 ± 3	3.4	62)
1,3-Dioxolane	$1.2 \cdot 10^{-2}$	15.6 ± 0.8	-15 ± 3	4.80	62)
	$9.42 \pm 13 \cdot 10^{-3}$	—	—	—	60)
	$7.71 \cdot 10^{-3}$ (23 °C)	15.6	-15	4.75	61)
2-Methyl-1,3-dioxolane	$7.9 \cdot 10^{-3}$	13 ± 1	-26 ± 4	4.85	62)
2-Phenyl-1,3-dioxolane	$1.5 \cdot 10^{-2}$	4 ± 1	-52 ± 3	—	62)
4,5-Dimethyl-1,3-dioxolane	$4.13 \cdot 10^{-2}$ (trans)	—	—	—	
	(23 °C)	12.6	-22.0	4.83	61)
	$7.35 \cdot 10^{-2}$ (cis) (23 °C)	11.7	-23.9	4.65, 4.95	61)
1,3-Dioxepane	$2.4 \div 3.0 \cdot 10^{-3}$ (22 °C)	—	—	—	
Dimethoxymethane	$6.0 \cdot 10^{-4}$	20 ± 2	-4 ± 6	4.50	62)

3.2.4 Alkoxycarbenium Ions

Olah prepared methoxycarbenium hexafluoroantimonate *22* by the reaction of CH_3OCH_2Cl with anhydrous $HF \cdot SbF_5$ in CH_2Cl_2[65]:

$$CH_3-O-CH_2Cl + HF \cdot SbF_5 \longrightarrow CH_3-O-\overset{+}{C}H_2\ SbF_6^- + HCl\uparrow \qquad (22)$$
$$22$$

The salt *22* precipitates from CH_2Cl_2 but is easily soluble in SO_2.

Recently, this salt (being a model of one of the possible isomeric forms of the active centers in polyacetals) was used to initiate the polymerization of 1,3-dioxolane and 1,3-dioxepane[66]. According to the ^{1}H- and ^{13}C-NMR spectra, the following species coexist in equilibrium at $-70\,°C$ when 1,3-dioxolane is used: (measurements were performed below monomer equilibrium concentration when polymerization was not possible[66]):

$$(23)$$

$$K_1 = \frac{[23a]}{[22]\,[DXL]} = 3 \cdot 10^3\ mole^{-1} \cdot l\ \text{and}\ \ K_2 = \frac{[23b]}{[23a]} = 3 \cdot 10^2$$

(First equilibrium was measured for dimethoxymethane and it is assumed that K_1 for DXL is roughly the same (cf. K_e^o for these two compounds in Table 3)).

It is remarkable that at $-70\,°C$ the exchange between the components of the equilibrium (*22*, DXL, *23a*, and *23b*) is sufficiently slow when compared with NMR time scale and that the component taken in excess could separately be observed (e.g. *22*). When *22* and DXL are used at the molar ratio 1:1, a simple ^{1}H-NMR spectrum is observed, consisting of two signals, namely $\delta = 4.25$ ppm (CH_3 and CH_2CH_2) and $\delta = 5.75$ ppm (OCH_2O) at a 7:4 integration ratio. ^{13}C-NMR gives three singlets: $\delta = 111.39$ ppm (CH_3), $\delta = 74.09$ ppm (CH_2CH_2) and $\delta = 64.86$ ppm (OCH_2O). According to the NMR spectra, the equilibrium is strongly shifted towards *23b*.

The reaction of *22* with the six-membered 1,3-dioxane (DXN) at low temperature gives mostly cationated DXN:

$$(24)$$

The transformation of the 6- to the more strained 8-membered ring by an analogy with DXL can only be observed at higher temperatures.

The seven-membered 1,3-dioxepane behaves in a similar way as DXN.

Alkoxycarbenium ions were shown to abstract hydride anions from the corresponding donors[67], but in the systems described above, practically no hydride transfer could be observed below $-30\,°C$.

It has also been shown earlier by Jaacks[68] that the methoxymethylium cation generated from the corresponding perchlorate (which probably exists in CH_2Cl_2 solution predominantly as an ester) is a very efficient initiator of DXL polymerization, and that only 1% of the cation participates at $0\,°C$ in the hydride transfer if DXL is used above its equilibrium concentration.

3.2.5 1,3-Dioxolan-2-ylium (Dioxolenium) Salts

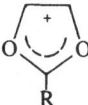

(where R = alkyl or H)

These salts can be considered as derivatives of dialkoxycarbenium ions.

As it was already discussed, these salts are easily formed by hydride transfer from the corresponding dioxolanes and triphenylmethylium salts but react differently with nucleophiles — exclusively by addition. Strong nucleophiles mostly form the kinetic product by substitution at C-2, weaker ones yield the thermodynamic products through ring opening at O-1—C-5 (O-3—C-4). These paths of the ambident reactivity, described in detail by Pittman[57] and by Perst[69], are shown below:

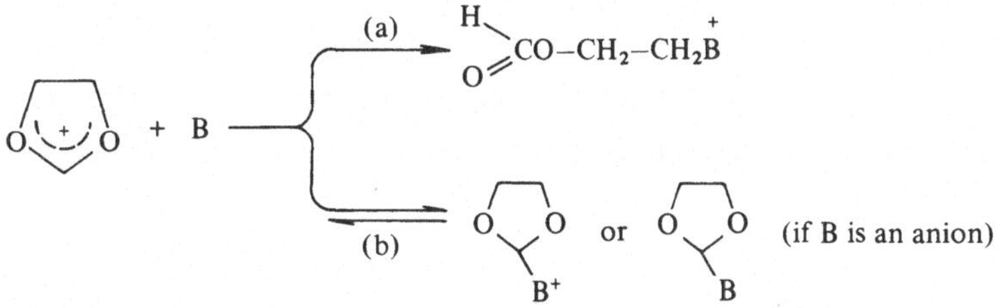

$$(25)$$

^{1}H-NMR spectra were used to identify structures of the first reaction products of THF[70] and DXL[71]. Since in both cases, the reaction proceeds according to route (a), the thermodynamic products are observed with characteristic ester groups

$$\left(\begin{matrix} O \\ \diagdown \\ H \end{matrix} C{-}O{-} ; \ ^1\text{H-NMR gives } \delta = 8.1 \text{ ppm for the formyl proton} \right)$$

1,3-Dioxolenium salts initiate polymerization of THF with rate constants comparable to that of propagation[71] and provide a class of convenient initiators that can be prepared with a variety of anions, since the starting triphenylmethylium salts are available with SbF_6^-, AsF_6^-, ClO_4^-, BF_4^-, and PF_6^- anions. A laboratory-scale preparation, identification, properties and ^{1}H-NMR spectra of unsubstituted dioxolenium salts are described in Refs. 71 and 72.

It has been also shown that the ^{1}H-NMR δ values of dioxolenium salts differ for salts with different anions and shift downfield for the free dioxolenium cation[63].

Yamashita[73] has used dioxolenium slats, prepared differently, as initiators for the polymerization of THF[73]. These salts yield dicationically growing chains whose

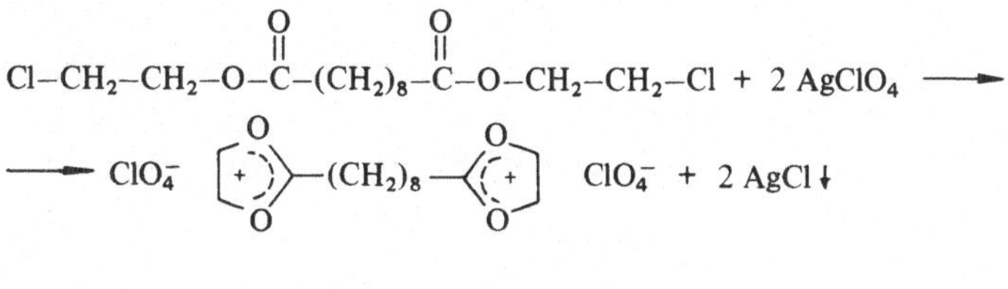

$$Cl{-}CH_2{-}CH_2{-}O{-}\overset{O}{\overset{\|}{C}}{-}(CH_2)_8{-}\overset{O}{\overset{\|}{C}}{-}O{-}CH_2{-}CH_2{-}Cl + 2\ AgClO_4 \longrightarrow$$

$$\longrightarrow ClO_4^- \ \left[\begin{matrix} + \\ O \end{matrix} \right]{-}(CH_2)_8{-}\left[\begin{matrix} + \\ O \end{matrix} \right] \ ClO_4^- + 2\ AgCl\downarrow \qquad (26)$$

structures were proved by formation of multi block copolymers with polystyrene/ α, ω-disodium alkanedicarboxylates.

Recently, an attempt was made to study the polymerization of various substituted DXL via the 1,3-dioxolenium initiation. However, these monomers gave substituted DXL salts failing to induce the polymerization of the parent monomers[61, 74].

3.2.6 Oxocarbenium Ions

These cations of the general formula

$$R-C\equiv\overset{+}{O} \; MtX_{n+1}^{-}$$

can easily be formed by mixing the corresponding acid halide with MtX_n (BF_3, PF_5, $SbCl_5$, SbF_5 etc.) or by using the silver salt technique:

$$RC\overset{O}{\underset{Y}{\diagdown}} + \; AgMtX_{n+1} \longrightarrow RC\equiv\overset{+}{O} \; MtX_{n+1}^{-} + \; AgY \downarrow \tag{27}$$

Both methods were used by Olah[75] who also showed by ^{1}H- and ^{19}F-NMR that, according to expectation, the corresponding salts exist in solutions in equilibria with their components:

$$R-C\overset{O}{\underset{X}{\diagdown}} + \; MtX_n \rightleftharpoons RC\equiv\overset{+}{O}MtX_{n+1}^{-} \tag{28}$$

Plesch[76] showed by electrochemical methods that at the sufficiently low concentrations, the components exist in the 2-to-2 equilibrium with unpaired ions[76].

These salts were used to initiate polymerization of cyclic ethers[77–79] and acetals[80, 81] but the structure of the first addition products has not yet been determined. Nevertheless, several authors observed the corresponding end groups in oligomers[78] (at the early stages of polymerization) or in high polymers[80]. These observations indicate that the oxocarbenium salts initiate by simple addition, for instance:

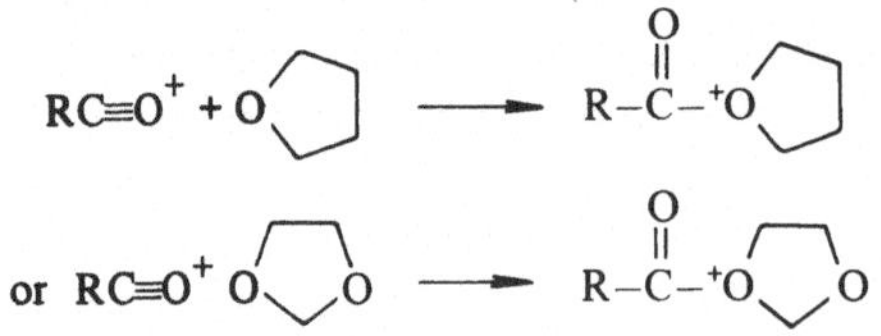

$$RC\equiv O^+ + O\bigcirc \longrightarrow R-\overset{O}{\underset{\|}{C}}-{}^+O\bigcirc$$

$$\text{or } RC\equiv O^+ \; O\overbigcirc O \longrightarrow R-\overset{O}{\underset{\|}{C}}-{}^+O\overbigcirc O \tag{29}$$

The ^{1}H-NMR spectrum of the reaction product of $CH_3CO^+SbF_6^-$ and THF in SO_2 has been reported although its resolution was poor and no assignements were given[78].

The presence of the C_6H_5-CO-O moiety in the polydioxolane chain was shown by ^{1}H-NMR and by UV. In order to detect the low concentration of the end groups in the ^{1}H-NMR spectrum, the polymer was prepared by Kubisa from the perdeuterated monomer[80]:

$$C_6H_5-\overset{\overset{\displaystyle O}{\|}}{C}-(O-CD_2-CD_2-O-CD_2)_n-X$$

Several multifunctional oxocarbenium salts were also prepared by Franta and used in the polymerization of THF[79]:

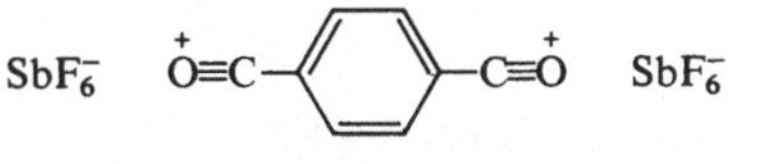

$$SbF_6^-\quad \overset{+}{O}{\equiv}C-\!\!\!\left\langle\!\!\!\bigcirc\!\!\!\right\rangle\!\!\!-C{\equiv}\overset{+}{O}\quad SbF_6^-\qquad SbF_6^-\quad \overset{+}{O}{\equiv}C-(CH_2)_4-C{\equiv}\overset{+}{O}\quad SbF_6^-$$

Terephthaloyl   Adipoyl

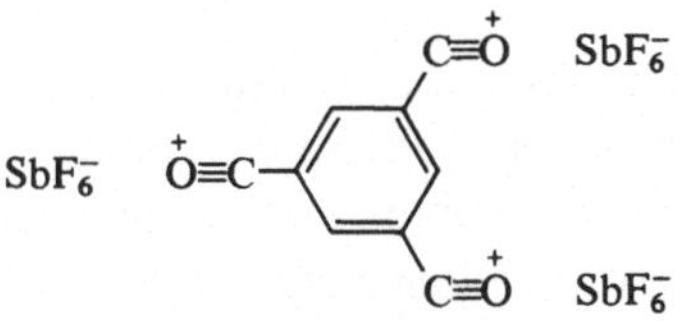

Trimesoyl

Initiation of THF with these salts gives indeed chains growing in two or three (trimesoyl) independent branches. This conclusion follows from $\overline{DP}_n$ measurements of isolated polymers and, moreover, from a comparison with the $\overline{DP}_n$ values of hydrolyzed polymers (Table 5).

Table 5. $\overline{DP}_n$ of Poly-THF prepared with terephthaloyl initiator and terminated by the LiBr[79]

Sample	Method of determination of $\overline{DP}_n$			Calc.
	UV	Br (detn.)	Osmometry	
1	319	331	340	355
2	230	266	245	255

Samples 1 and 2 of Table 5 were hydrolyzed (ester bonds joining the polyether chain with the initiator); after hydrolysis the $\overline{DP}_n$ value measured by osmometry was half of that measured before hydrolysis.

Kinetic measurements, assuming propagation from two ends, gave efficiencies of initiation equal to 0.93 and 0.96 for samples 1 and 2, respectively.

Although no systematic study is available for the rates of initiation, it seems that, on the basis of the preliminary comparison of the ^{1}H-NMR spectra with the kinetics of polymerization (proceeding with an autoacceleration period), the first addition is fast and practically irreversible ($k_{i1} \gg k_{-i1}$):

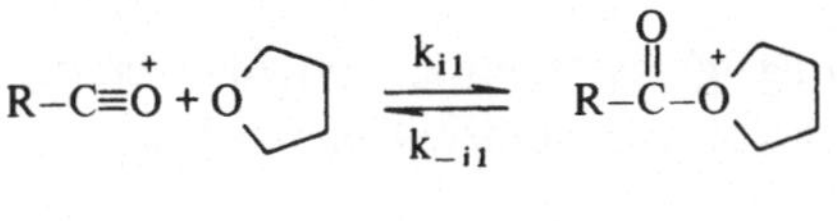

$$R-C{\equiv}\overset{+}{O} + O{\bigcirc} \underset{k_{-i1}}{\overset{k_{i1}}{\rightleftharpoons}} R-\overset{\overset{\displaystyle O}{\|}}{C}-\overset{+}{O}{\bigcirc} \tag{30}$$

whereas the addition of the second monomer molecule is slow ($k_{i1} \gg k_{i2} < k_p$):

$$R-\overset{\overset{O}{\|}}{C}-\overset{+}{O}\bigcirc \;+\; O\bigcirc \;\underset{\phantom{k_{i2}}}{\overset{k_{i2}}{\rightleftharpoons}}\; R-\overset{\overset{O}{\|}}{C}-O-(CH_2)_4-\overset{+}{O}\bigcirc \tag{31}$$

3.2.7 Onium Ions

Oxonium, sulfonium and ammonium ions will be described in this section:

$$R_3O^+A^- \quad R_3S^+A^- \quad \text{and} \quad R_4N^+A^- \quad (\text{where } A^- = MtX_{n+1}^-)$$

The synthesis of the most versatile trialkyloxonium ions, which can initiate all of the considered classes of heterocycles (cyclic acetals, ethers, sulfides, lactones, phosphates, amines) is simple in the case of BF_4^- and $SbCl_6^-$ anions[82, 83]. It involves reaction of an ether with α-epichlorohydrin and the corresponding metal halide, e.g.:

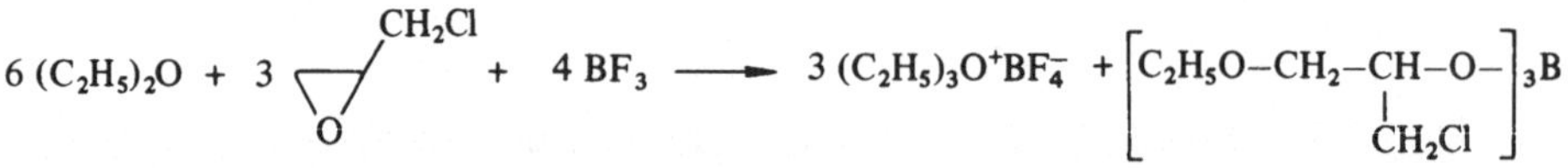

$$6\,(C_2H_5)_2O \;+\; 3\;\overset{CH_2Cl}{\underset{O}{\triangle}} \;+\; 4\,BF_3 \;\longrightarrow\; 3\,(C_2H_5)_3O^+BF_4^- \;+\; \left[C_2H_5O-CH_2-\underset{\underset{CH_2Cl}{|}}{CH}-O-\right]_3B \tag{32}$$

Trialkyloxonium salts with the more stable PF_6^- or SbF_6^- anions can be prepared, according to Olah, by using orthoformates as the alkylating agents[84]:

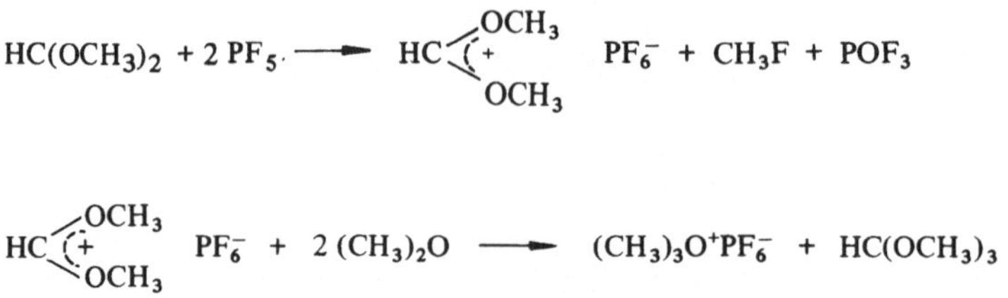

$$HC(OCH_3)_2 + 2\,PF_5 \longrightarrow HC\overset{OCH_3}{\underset{OCH_3}{\overset{+}{\diagdown}}}\; PF_6^- + CH_3F + POF_3 \tag{33}$$

$$HC\overset{OCH_3}{\underset{OCH_3}{\overset{+}{\diagdown}}}\; PF_6^- + 2\,(CH_3)_2O \longrightarrow (CH_3)_3O^+PF_6^- + HC(OCH_3)_3$$

Recently, another method, which seems to be the most general one, was described[85]:

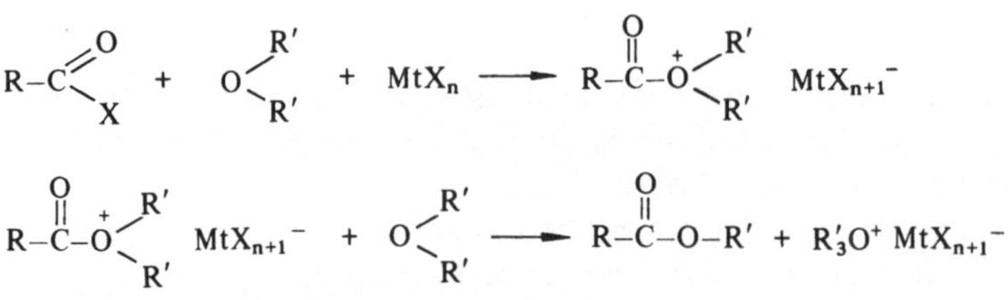

$$R-C\overset{O}{\underset{X}{\diagdown}} \;+\; O\overset{R'}{\underset{R'}{\diagup}} \;+\; MtX_n \longrightarrow R-\overset{\overset{O}{\|}}{C}-\overset{+}{O}\overset{R'}{\underset{R'}{\diagup}}\; MtX_{n+1}^- \tag{34}$$

$$R-\overset{\overset{O}{\|}}{C}-\overset{+}{O}\overset{R'}{\underset{R'}{\diagup}}\; MtX_{n+1}^- \;+\; O\overset{R'}{\underset{R'}{\diagup}} \longrightarrow R-\overset{\overset{O}{\|}}{C}-O-R' \;+\; R'_3O^+\,MtX_{n+1}^-$$

The preparation of $(C_2H_5)_3O^+\,SbF_6^-$ and oxonium salts with higher alkyl groups can readily by achieved by a one-step synthesis.

The preparation of the oxonium salts with other anions should also be possible this way.

Trialkyloxonium salts being strong alkylating agents[86], initiate polymerization by simple alkylation at the most nucleophilic site of a monomer. This was shown for many organic bases and also observed for heterocyclic monomers. Saegusa studied the initiation of THF by $(C_2H_5)_3O^+BF_4^-$ and directly observed by ^{1}H-NMR the formation of the ethylated THF (ethyltetrahydrofuranium cation) and $(C_2H_5)_2O$[87]:

$$(C_2H_5)_3O^+BF_4^- + O \rightleftharpoons C_2H_5-\overset{+}{O} \quad BF_4^- + (C_2H_5)_2O \qquad (35)$$

This first addition reaction was also studied for the polymerization of THF initiated with $(CH_3)_3O^+SbF_6^-$. The corresponding methyltetrahydrofuranium cation:

$$CH_3-\overset{+}{O} \quad SbF_6^-$$

was identified on the basis of its ^{1}H-NMR spectrum and isolated as a pure crystalline compound (m.p. = 178 °C)[88].

A simple alkylation, without any side reactions was also observed in the initiation of cyclic sulfides[89] and of cyclic esters of phosphoric acid 2-methoxy-2-oxo-1,3,2-dioxaphosphorinane)[90]. In the latter case, the tetraalkoxyphosphonium ion is formed (the phosphoryl oxygen atom is alkylated) as follows:

$$(C_2H_5)_3O^+BF_4^- + S \underset{CH_3}{\overset{CH_3}{\diamondsuit}} \xrightarrow[-(C_2H_5)_2O]{} C_2H_5-\overset{+}{S} \underset{CH_3}{\overset{CH_3}{\diamondsuit}} \quad BF_4^- \qquad (36)$$

$$(C_2H_5)_3O^+SbF_6^- + \underset{\overset{P}{\diagdown}}{O \quad O} \xrightarrow[-(C_2H_5)_2O]{} \underset{C_2H_5O \quad OCH_3}{O \quad \overset{+}{P} \quad O} \quad SbF_6^- \qquad (37)$$

The polymerization of cyclic acetals was also initiated by trialkyloxonium salts but the first addition product, a cationated cyclic acetal, could neither be isolated nor directly observed. This is mainly due to the much faster subsequent propagation. Nevertheless, initiation by addition was detected by studies of the end groups of oligomers of high polymers. Ponomarenko and Ludvig[91] used the ^{14}C labelled triethyloxonium salt to initiate polymerization of 1,3-dioxolane, and quantitative initiation was found for the equilibrated system[91]. The same conclusion was reached from FT-^{1}H-NMR studies of the end groups of poly-1,3-dioxolane prepared by Kubisa[92] from perdeuterated 1,3-dioxolane-d$_6$[92]:

$$CH_3CH_2-(O-CD_2-CD_2-O-CD_2)_n-X$$

The latter method has allowed the quantitative determination of the concentration of ethoxy end groups in polymers with $\overline{DP}_n$ as high as 10^3. Small and precisely known proportions of the non-deuterated mers serve as internal standard. This work also indicates that, with the properly chosen conditions (when the proportion of macrocycles is minimized) and with well purified components, one molecule of initiator produces one macromolecule.

There is also one report available claiming that $(C_2H_5)_3O^+BF_4^-$, reacting with 1,3-dioxolane, produces some ethane as a side product through hydride abstraction. In the same work no hydride transfer was observed for $(C_2H_5)_3O^+SbF_6^-$[93].

It seems thus rather well documented that trialkyloxonium salts initiate polymerization of heterocyclic monomers by simple addition and without side reactions, at least when stable anions are used.

The other onium salts (sulfonium or ammonium salts) are less often applied due to their lower reactivity. They can neither initiate the polymerization of cyclic ethers or acetals nor be displaced by less nucleophilic ligands (cyclic sulfides and amines respectively).

Generally, the position of the equilibrium in reaction (38) depends on the nucleophilicity of centers X and Y:

$$\underset{\diagdown}{\overset{\diagup}{X}}{\overset{+}{-}} + \bigcirc\!Y \ \underset{}{\overset{K}{\rightleftharpoons}}\ \underset{\diagdown}{\overset{\diagup}{X}} + \ -\overset{+}{Y}\bigcirc \ \underset{\text{Propagation}}{\rightleftharpoons} \tag{38}$$

Thus, halonium ions (e.g. Et_2Cl^+, A^-) can initiate polymerization of acetals, ethers, sulfides, or amines. Oxonium ions also initiate more or less successfully the polymerization of these monomers while ammonium ions can only be applied in the polymerization of some cyclic amines.

The utility of onium ions depends not only on the value of K and related rate constants of the equilibration involved in initiation but also on the rate constant of the subsequent propagation. Thus, in spite of the much higher nucleophilicity of ethers, cyclic acetals can be polymerized by trialkyloxonium salts, because of the fast propagation. Therefore, $-\overset{+}{Y}\bigcirc$ species, formed in the first equilibrium, are continously removed and even quantitative initiation is possible in spite of the low K values (e.g. $K < 10^{-1}$).

Detailed studies were performed for a series of cyclic ethers and K values measured for $(C_2H_5)_3O^+BF_4^-$ and THF (K = 36 at 35 °C in CH_2Cl_2), THP (K = 23.4) and OXP (K = 26)[87]. Cyclic ethers are thus not only more basic but also more nucleophilic (reaction (38)) than linear ethers.

Cyclic sulfides are known to be less basic than linear sulfides. Because the equilibrium constant K (Eq. (38)) is small, trialkylsulfonium ions cannot be used as initiators for the polymerization of cyclic sulfides[94]. The relation between cyclic and linear amines resembles that of sulfides rather than ethers. In some cases, however, ammonium ions have been successfully used as initiators, e. g. in the polymerization of conidine[40] or 1-benzylaziridine[95].

3.2.8 Covalent Initiators ("Hidden Cations")

Strong alkylating compounds (through cationation) belong to this category of initiators. The possibility to cationate depends not only on the inherent cationating ability but also on the nucleophilicity of a given monomer.

Thus, one could, in principle, construct the scale of cationating power and nucleophilicities from which it would be possible to conclude a priori whether polymerization of a given monomer can be initiated by a chosen covalent initiator.

At the top of the ladder of the cationating power, there are esters of strong protonic acids (superacids) like CF_3SO_3R, FSO_3R, $ClSO_3R$, etc.

As is kown for other S_N2 reactions[97] the cationating power increases with decreasing size of the ester group $(CH_3 > C_2H_5 > C_3H_7)$[96]. On the other hand, even the weakest cationating agents are able to initiate the polymerization of strongly nucleophilic monomers like aziridines, oxazolines etc.

Esters of strong acids can be used for all classes of heterocyclic monomers. Initiation, in all cases studied involves simple cationation; it usually proceeds less rapidly than cationation with trialkyloxonium salts. No side reactions were reported, although macroesters can be formed, due to the possibility of collapse of the ion pairs formed in the first step (cf. Chap. 5).

Initiation of THF with ethyl triflate leads to ethylated THF observed directly by ^{1}H-NMR[98]:

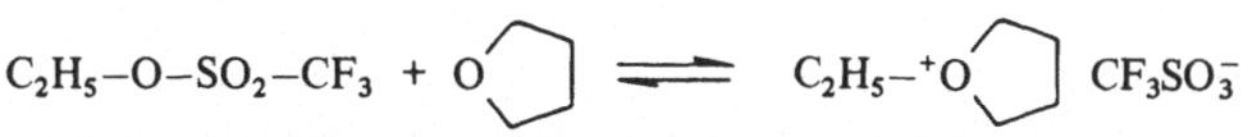

$$C_2H_5-O-SO_2-CF_3 + O\bigcirc \rightleftharpoons C_2H_5-^+O\bigcirc \quad CF_3SO_3^- \qquad (39)$$

Because the triflic anion is less nucleophilic than THF, the ion pair so formed propagates before collapsing to the corresponding ester.

Initiation of the polymerization of 2-oxazolines (much more nucleophilic than cyclic ethers) is possible with much weaker alkylating agents[99]. It has been shown that e.g. benzyl bromide can be used:

$$C_6H_5-CH_2Br + N\bigcirc O \rightleftharpoons C_6H_5-CH_2-N\overset{+}{\bigcirc}O \quad Br^- \qquad (40)$$
$$\qquad\qquad R \qquad\qquad\qquad R$$
$$\qquad\qquad\qquad\qquad\qquad 40$$

The oxazolinium salts _40_ (isomeric form of the corresponding tertiary oxonium salt) formed according to Eq. (40) are in equilibrium (because of the high nucleophilicity of the bromide anion) with the corresponding alkyl bromide:

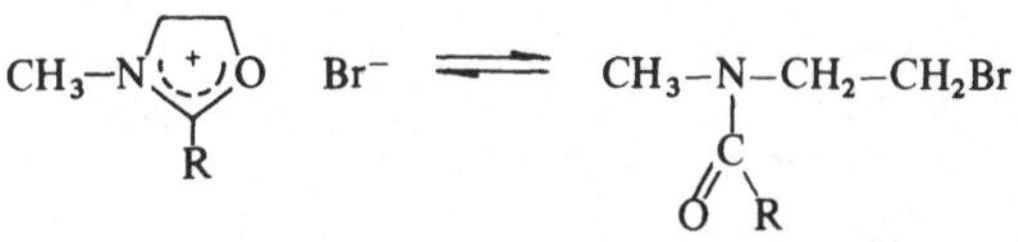

$$CH_3-N\overset{+}{\bigcirc}O \quad Br^- \rightleftharpoons CH_3-N-CH_2-CH_2Br \qquad (41)$$
$$\qquad R \qquad\qquad\qquad\qquad \overset{C}{\underset{O}{\Big|}} \;\; R$$

This possibility of the ion pair to collapse, proved for several systems, was not considered in the polymerization of bicyclic amines initiated for instance with ethyl bromide. The assumed structure of the cationated conidine is[40]:

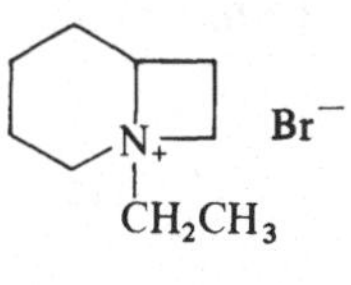

It has to be studied whether the corresponding alkyl bromide (in analogy with 2-oxazolines) does exist in equilibrium with the ammonium salt under the polymerization conditions.

Anhydrides of strong protonic acids provide a group of initiators able to give dicationically terminated macromolecules[100]. The anhydride of trifluoromethanesulfonic acid (triflic anhydride) initiates the polymerization of THF in this way; both reactions, with rate constants k_1 and k_{21}, are faster than the formation of the alkyltetrahydrofuranium cation with the corresponding triflic acid ester:

$$CF_3-SO_2\diagdown_{} O + \diagup_{O} \xrightarrow{k_1} CF_3-SO_2-O\overset{+}{\diagup_{O}} \quad CF_3-SO_3^- \xrightarrow{k_{21}} \diagup_{O}$$

$$CF_3-SO_2\diagup$$

$$(42)$$

$$\longrightarrow CF_3-SO_2-O-(CH_2)_4-\overset{+}{O}\diagup_{} \quad CF_3-SO_3^-$$

Neither the first nor the second step in Scheme (42) has been directly observed, but it has conclusively been shown by studies of the number of the end groups in poly-THF initiated with triflic anhydride that one molecule of the anhydride gives two growing ends of the cationic and/or ester structure in one macromolecule. These structures are interconvertible. Recently, Smith[101] separated and characterized two intermediate products: tetramethylene-bis-triflate, resulting from the attack of the anion on the α-methylene carbon atom in salt *42a*, and the corresponding dicationic trimer

$$\diagup_{O^+}-(CH_2)_4-\overset{+}{O}\diagup_{} \cdot 2\ CF_3SO_3^-$$

The diester persits in solution because it can only be converted into the ion pair by slow bimolecular reaction with THF; the faster intramolecular ionization is not possible because of the absence of nucleophilic oxygen atoms. The corresponding dicationic trimer reacts with THF with a rate constant comparable to k_p.

3.2.9 Friedel-Crafts (F-C) Initiators

Lewis acids have been successfully used as initiators for the cationic polymerization of various heterocycles. Although BF_3 complexes are routinely used to check the polymerizability of newly synthesized monomers and in the preparation of new polymers, the chemical reactions involved in initiation are not known well enough for many hundreds of systems in which BF_3 or other F-C initiators have been used. Chemical reactions preceding the actual initiation step may markedly depend on the structure of the monomer involved and of the F-C initiator. Thus, in the polymerization of vinyl compounds (at least for isobutene) in hydrocarbon solvents of low polarity, the self-ionization of some Lewis acids probably precedes the formation of the first propagating species[102], e.g.:

$$2\ AlBr_3 \rightleftharpoons AlBr_2^+,\ AlBr_4^- \tag{43}$$

The preformed cation can then be added to the monomer molecule.

In the polymerization of heterocycles such a possibility can be omitted because Lewis acids like BF_3 readily form complexes with all nucleophilic reagents such as ethers, ketones, sulfides, or amines.

The equilibrium constant for the formation of e.g. the BF_3/Et_2O complex is equal to $K = 140$ mole$^{-1} \cdot l$ at 25 °C[103]. This value is even larger for more nucleophilic monomers. BF_3 complexes with the majority of ethers are sufficiently stable to be distilled without decomposition (e.g. $BF_3 \cdot THF$ complex: b.p.$_{20} = 98.5$ °C)[104]. This complex exchanges THF with an excess of this monomer present in the system:

$$F_3B \cdot O\bigcirc + O\bigcirc \underset{}{\overset{k_1}{\rightleftharpoons}} F_3B \cdot O\bigcirc + O\bigcirc \qquad (44)$$

The exchange reaction was studied, for instance, in the model system $Et_2O \cdot BF_3/THF$ using ^{19}F and ^1H-NMR methods:

$$F_3B \cdot O{\overset{C_2H_5}{\underset{C_2H_5}{<}}} + O\bigcirc \underset{k_{-1e}}{\overset{k_{1e}}{\rightleftharpoons}} F_3B \cdot O\bigcirc + O{\overset{C_2H_5}{\underset{C_2H_5}{<}}} \qquad (45)$$

Values of the rate constants k_{1e} and k_{-1e} were found to be $k_{1e} = 9 \cdot 10^2$ mole$^{-1} \cdot$ l $\cdot$ s^{-1}, $k_{-1e} = 2 \cdot 10^2$ mole$^{-1} \cdot$ l $\cdot$ s^{-1} at 25 °C in CH_2Cl_2[105].

The exchange between two complex molecules, e. g.:

$$Et_2O \cdot BF_3 + \bigcirc O \cdot BF_3 \underset{}{\overset{k_2}{\rightleftharpoons}} \bigcirc O \cdot BF_3 + Et_2O \cdot BF_3 \qquad (46)$$

was estimated to be much slower ($k_2 < 7$ mole$^{-1} \cdot$ l $\cdot$ s^{-1})[105]. The enthalpies of activation were also calculated ($\Delta H^{\ddagger}_{1e} = 13$ kcal $\cdot$ mole^{-1}, $\Delta H^{\ddagger}_{-1e} = 11.4$ kcal $\cdot$ mole^{-1} and $\Delta H^{\ddagger}_2 = 19$ kcal $\cdot$ mole^{-1}). The exchange between free BF_3 (taken in excess) and its complex with THF proceeds as a bimolecular reaction with the rate constant $k_3 = 4 \cdot 10^3$ mole$^{-1} \cdot$ l $\cdot$ s^{-1} for THF at 25 °C in toluene ($\Delta H^{\ddagger}_3 = 4.6$ kcal $\cdot$ mole^{-1})[106]:

$$BF_3 + \bigcirc O \cdot BF_3 \underset{}{\overset{k_3}{\rightleftharpoons}} BF_3 + \bigcirc O \cdot BF_3 \qquad (47)$$

Self-dissociation of the complex is of less importance.

Recently, the ^{13}C-NMR method was also applied to these studies and the following order of the relative strengths of Lewis acids and bases, related to equilibrium (47) was given:

$$BBr_3 > BCl_3 > BF_3 > BH_3 \text{[107]}; \qquad THF > Et_2O > Pr_2O > Bu_2O \text{[108]}$$

In spite of numerous applications of BF_3 and related compounds as initiators, the involved reactions have not been studied in detail.

In a series of papers Korovina and Entelis have proposed the zwitterionic polymerization of THF initiated by BF_3 in the presence of propylene oxide[111, 112]. Unfortunately, the collected evidence is of indirect nature. Recently, according to Entelis[113], the ^1H-NMR and ^{19}F-NMR methods have confirmed the earlier proposals[113].

The mechanism of initiation was carefully studied for the PF_5-THF system[109, 110]. Reichert[109] used the ^{31}P-NMR method and claimed the following set of reactions[109]:

$$2 F_5P \cdot O\bigcirc \longrightarrow F_4PO-(CH_2)_4-\overset{+}{O}\bigcirc , PF_6^- \xrightarrow[-POF_3]{} F-(CH_2)_4-\overset{+}{O}\bigcirc , PF_6^- \xrightarrow[\text{Chain growth}]{+ O\bigcirc}$$

$$\downarrow F_5P \cdot O\bigcirc$$

$$\text{Dicationic chain growth} \xleftarrow{+ O\bigcirc} PF_6^- , \bigcirc \overset{+}{O}-(CH_2)_4-\overset{+}{O}\bigcirc , PF_6^-$$

Only two signals were observed in the ^{31}P-NMR spectrum, namely that of the $F_5P \cdot THF$ complex (δ 134.9 ppm) and of the PF_6^- anion (δ 145 ppm). PF_3O (b.p. = −40 °C) was detected by IR in the gaseous phase. End groups $F_4PO\sim$ were not observed. To explain their absence, it was proposed that these end groups rapidly decompose to fluoroalkyl end groups which, in turn, were also not detected. A weak point of this mechanism is the assumption that the fluoroalkyl end groups are formed because they are known to react slowly with any Lewis acid. It seems to be more realistic to suppose that the bimolecular reaction between the $F_4PO^\sim$ group and a new complex molecule occurs in one step only.

There is one more paper on the same subject based on ^{19}F-NMR studies. In this work, the formation of some side products (e. g. $F-(CH_2)_4-F$) had to be proposed but the postulated compounds have not been observed by ^{19}F-NMR [110].

An apparent controversy exists between the known high stability of complexes of BF_3 or PF_5 with ethers and the mechanism of initiation postulated in Scheme (48). Indeed, the F_5P : THF complex (m.p. = 55 °C, b.p.$_{0.15}$ = 116 °C) [114] can even be sublimed at 70 °C/0.02 mm Hg. This reflects its inherent (thermal) stability when no nucleophiles are present. However, in the presence of an excess of THF, the reactions described by Scheme (48) start to proceed, because excess of THF provides stabilization of the cation formed.

The ability of Lewis acids to initiate polymerization without any cocatalyst added (as in Eq. (48)) depends on both the ring strain and the nucleophilicity of monomer as well as on the Lewis acid strength. For example, while BF_3 cannot initiate the polymerization of THF (ring strain ~ 5.0 kcal $\cdot$ mole^{-1}), it does initiate the polymerization of the three-membered oxirane (ring strain ~ 20 kcal $\cdot$ mole^{-1}) giving a complex with BF_3, which is only stable below -80 °C. It was also shown that BF_3 does not initiate the polymerization of some other monomers including oxetane [115], BCMO [116] and propylene sulfide [41]. Polymerization can be started in these systems by the addition of proton donors or other promoters forming ions with BF_3.

In a few systems studied, the rate of polymerization first increases up to a 1 : 1 ratio of $H_2O : BF_3$ and then tends to decrease with further addition of water (from this dependence, the concentration of the residual water could be determined by extrapolation to zero rate). In all these systems, the complex acid "$H^+BF_3OH^-$" was proposed as the initiator. In this formula, the proton is tacitly assumed to be solvated by (at least) monomer molecules or exists as a secondary oxonium ion (cf. Sect. 3.1).

Initiation of some monomers (e. g. THF) by the Lewis acid $-$ H_2O system is often so slow that the addition of so-called promoters is required (oxiranes [111, 117], lactones [117]). First, these promoters rapidly form the secondary oxonium ions replacing the slow initiation with the less reactive monomer (here, less strained THF).

Thus, Lewis acids do not appear to be prominent initiators for living polymerizations. Initiation is usually slow, efficiency is low and side products (e.g. POF_3, reactive end groups) could induce termination or undesirable transfer reactions.

3.2.10 Cations Formed by Electron Transfer

The conversion of free radicals by electron transfer into carbocations and subsequent initiation of the cationic polymerization has recently been reviewed [118]. The electron transfer process

$$R^\cdot \xrightarrow{-e} R^+ \qquad\qquad\qquad (49)$$

is favoured by the presence of electron-releasing substituents, attached to the carbon-centered radical, stabilizing the carbocation and thus lowering both the ionization potential and the oxidation potential of the free radical [118]. It is important to use the suitable oxidant which simultaneously supplies the necessary anion.

Recently, two groups of systems were investigated in greater detail using THF as a model monomer. In the first group, aryldiazonium salts with stable anions (e. g. PF_6^-) were employed together with thermal or photochemical sources of free radicals. In the photochemically induced formation of the initiating cation, the following sequence of reactions was proposed for 2,2-dimethoxy-2-phenylacetophenone as the photoactive radical source[119].

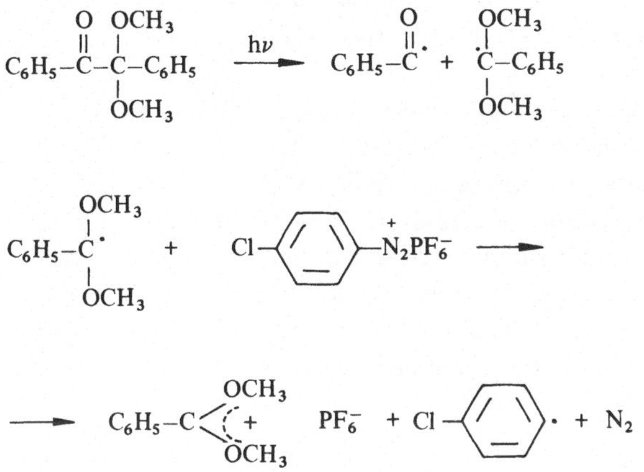

(50)

This system initiates polymerization of THF giving, upon irradiation, a high polymer ($\overline{M}_n \simeq 5.0 \cdot 10^4$).

In the second group of systems diarylhalonium salts, failing to initiate polymerization generate the carbenium salts by decomposition induced by free radicals formed photochemically. Benzophenone, benzil or maleic anhydride were used as sources of free radicals[119]:

$$R_2 \xrightarrow{h\nu} 2R\bullet \qquad\qquad (51)$$

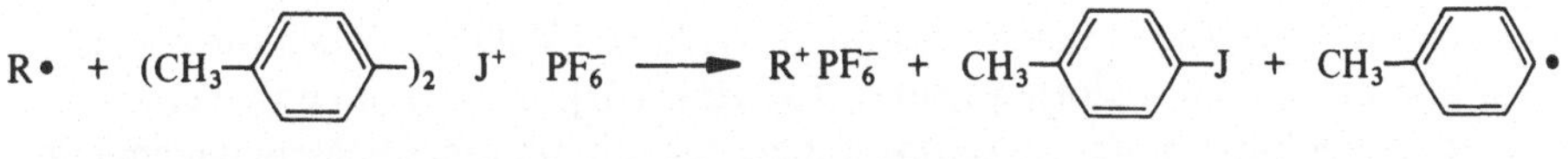

Depending on the structure of the carbocation formed and monomer used, initiation can proceed either by hydride abstraction[118] or by direct addition.

The photochemical decomposition of sulfonium or diarylhalonium salts can also lead to the formation of protonic acids subsequently initiating polymerization[120, 121]:

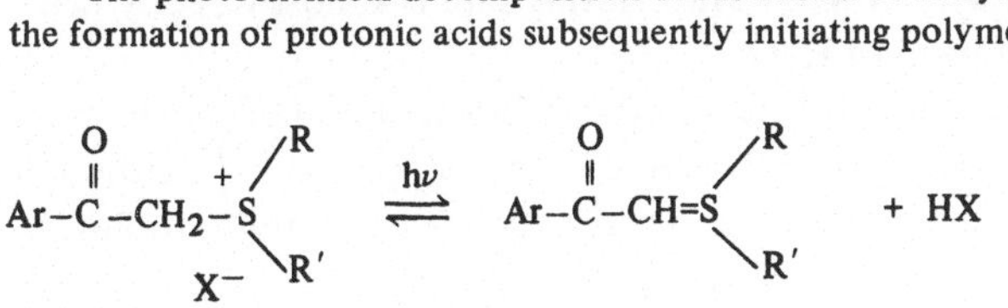

(52)

or

$$Ar_2I^+ A^- \xrightarrow{h\nu} [Ar_2I^+ A^-]^* \longrightarrow Ar-I^+ + Ar^\cdot + A^-$$

$$Ar-I^+ + S-H \longrightarrow Ar-I^+H + S^\cdot \qquad\qquad (53)$$

$$Ar-I^+ -H \longrightarrow Ar-I + H^+$$

Thus, the principle of the formation of carbocations, clearly shown by Scheme (51), requires judicious choice of both components: the source of free radicals should not only be able to dissociate easily — either thermally or photochemically — but also exhibit a chemical structure which stabilizes the resultant carbenium ion (e.g. the methoxy groups and an aromatic ring as in Scheme (50). On the other hand, the cationic component should provide, preferably in the radical-induced reaction, a cation which abstracts an electron from the free radical (Scheme 50).

From time to time it is claimed that the polymerization of THF can be initiated by some unusual processes or unusual initiators. For instance, it has recently been reported that silver and copper salts with anions like AsF_6^-, BF_4^-, SbF_6^-, $SbCl_6^-$, $CF_3SO_3^-$ etc. initiated the polymerization of THF at 25 °C when irradiated during a prolonged time[42]. Obviously, countless number of systems can be imagined, where drastic conditions will destroy the compounds used and produce, for instance, an usual initiator like protonic acid. It would be advisable to verify this simplest assumption before any more complicated mechanisms are proposed.

3.2.11 Organometallic Initiators

This group usually leads to anionic or coordinate polymerization which are not covered by this review. Nevertheless, polymerizations of oxetanes and THF, known to proceed exclusively by a cationic mechanism, have also been induced by various organometallic initiators. In many cases, these initiators lead to higher molecular weight polymers, probably reacting fast and first with impurities that could, if not destroyed, lower the molecular weight by chain transfer.

Aluminum alkyl initiators usually need water to be converted into their active form (alumoxanes). This has clearly been shown in the polymerization of 3,3-bis-(chloromethyl)-oxetane initiated with $(i\text{-}C_4H_9)_3Al$ and H_2O[122]. When the overall rate of polymerization is plotted against $[(i\text{-}C_4H_9)_3Al]_0$, there is no polymerization observed at low initiator concentrations; then, the rate becomes linearly dependent on concentration and eventually independent. These dependences for two different initial concentrations of H_2O added to the system ($5.25 \cdot 10^{-3}$ mole $\cdot$ l^{-1} and $1.25 \cdot 10^{-2}$ mole $\cdot$ l^{-1}) are shown in Fig. 2. The limiting rate is almost directly proportional to the initial concentration of H_2O.

These dependences were further confirmed in the polymerization of 3-chloromethyl-3-methyloxetane studied by Aleksiuk, Alferova and Kropachev[123], although these authors observed the maximum rate of polymerization for different ratio of H_2O to aluminum alkyls.

Two of us proposed[122] that in the reaction of $(i\text{-}C_4H_9)_3Al$ with H_2O, the active initiator is formed and that for every active site approx. ten molecules of H_2O are used. This latter proposal came from comparison of DP_n of resulting polymers and starting [monomer]/[water] ratio. Thus, the initiating species has the structure:

$$\diagdown Al-O-(Al-O-)_n \cdots$$

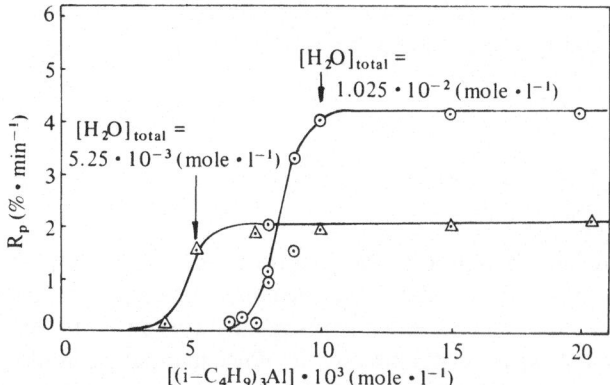

Fig. 2. Dependence of the rate of polymerization R_p of BCMO on $[(i-C_4H_9)_3Al]$ for two different concentrations of water. $[M]_0 = 0.15$ mole $\cdot$ l^{-1}, 70 °C, C_6H_5Cl solvent (Ref. 122)

The counterion is very stable; in the polymerization of 3,3-bis-(chloromethyl)-oxetane no termination, due to its decomposition, was observed up to 90 °C. Thus, it was proposed that the anion has a stabilized cyclic structure, formed after preliminary coordination of monomer with initiator:

$$\tag{54}$$

The first step of coordination may proceed with a rate comparable with that of dissolution, then, the rapidly formed addition product reacts slower with a next monomer molecule and ionizes in the following way:

$$\tag{55}$$

Since the polymerizations of 3,3-bis-(chloromethyl)-oxetane[122] and 3-chloromethyl-3-methyl-oxetane[123] were studied in chlorobenzene, a non-polar solvent, both ends of the macromolecule can constitute one ion pair, at least at the earlier stages of polymerization.

The mechanism described above is based on all kinetic information available but the proposed species have not directly been observed.

3.3 Kinetics of Initiation

The initiation process may involve either a single or a few consecutive elementary reactions. In the latter case, the understanding of this process requires knowledge of all these steps: their chemistry and rate constants. NMR spectroscopy proved to be the experimental method of choice in kinetic and mechanistic studies of initiation. However, UV spectroscopy, polarography, conductivity, and the ^{14}C tracer method were also applied. Because of the complexity of initiation, the kinetic studies were performed only for a limited number of systems. The rate constants of the elementary reactions were only determined for a few simpler systems.

Let us assume that initiation in Scheme (56) is irreversible and that the rate constant of the addition of the first monomer molecule to the active species (k_{p1}) is either equal to or higher than the rate constant of propagation with an active center bound to a polymer molecule of any length ($k_{p1} \leqslant k_{pn}$).

$$I + M \xrightarrow{\ k_i\ } P_1^x$$

$$P_1^x + M \xrightarrow{\ k_{p1}\ } P_2^x \quad \text{where} \ \ k_{p1} \geqslant k_p \qquad\qquad\qquad (56)$$

$$P_n^x + M \xrightarrow{\ k_p\ } P_{n+1}^x$$

The solution of this scheme gives:

$$\ln \frac{[I]_0}{[I]_t} = k_i \int_0^t [M]\, dt , \qquad\qquad\qquad\qquad (57)$$

which can be simplified to Eq. (58) if propagation is much slower than initiation:

$$\ln \frac{[I]_0}{[I]_t} = k_i [M]_0 \cdot t \qquad\qquad\qquad\qquad (58)$$

(provided $[M]_t \approx \text{const} \approx [M]_0$)

Thus, measuring the concentration of initiator as a function of time, the rate constant of initiation can be determined. In a few systems, when the equilibrium monomer concentration is relatively high (THF, DXL), initiation has been studied under conditions at which propagation is excluded ($[M]_0 < [M]_e$).

For example, when ethyl triflate ($C_2H_5OSO_2CF_3$) is reacted with THF in CD_3NO_2 solvent at 25 °C at $[I]_0 = [M]_0 = 2.0$ mole $\cdot$ 1^{-1}, only the formation of the ethyltetrahydrofuranium ion is observed without any subsequent propagation[98]:

$$CH_3-CH_2-O-SO_2-CF_3 \; + \; O\underset{CH_2-CH_2}{\overset{CH_2-CH_2}{\diagup}} \; \rightleftharpoons \; CH_3-CH_2-\overset{+}{O}\underset{CH_2-CH_2}{\overset{CH_2-CH_2}{\diagup}} \; , \; {}^-O-SO_2-CF_3$$

a   b   A   c   d   e   f   g   h   B

$$(59)$$

^{1}H-NMR spectra of this system, recorded after different times from the beginning of the reaction, are shown in Fig. 3. Reaction (59) can also be followed by ^{19}F-NMR, recording signals A ($\delta = 88.0$ ppm from C_6F_6) and B ($\delta = 85.2$ ppm). The reaction is treated as a second-order reaction and the plot of $[THF]_t^{-1}$ vs. time gives the rate constant of initiation $k_i = 5 \cdot 10^{-4}$ mole$^{-1} \cdot 1 \cdot s^{-1}$.

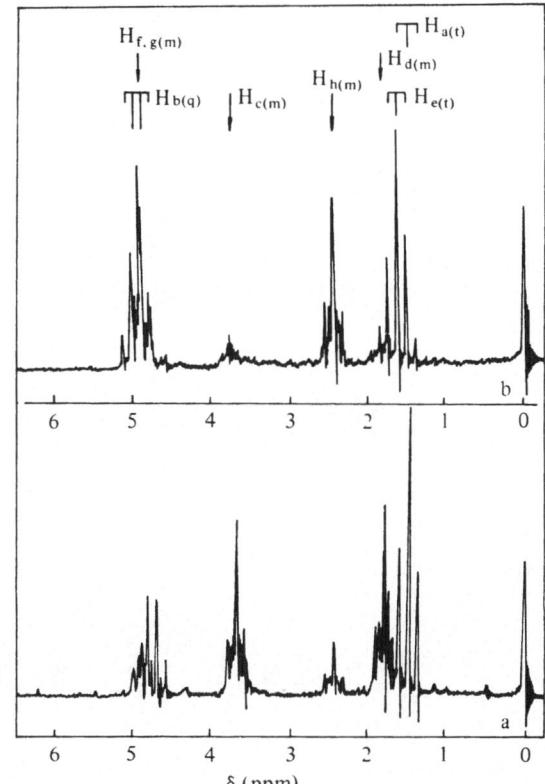

Fig. 3. Reaction (59) studied by ^{1}H-NMR. Spectra of the reaction mixture after 10 min (a) and 60 min (b). $[C_2H_5OSO_2CF_3]_0 = 2.0$ mole $\cdot$ l^{-1}, $[THF]_0 = 2.0$ mole $\cdot$ l^{-1}, CD$_3$NO$_2$ solvent, 25 °C

δ (ppm)

Reactions of 1,3-dioxolenium salts with THF and DXL were studied in a similar way. It was found from the NMR spectra that initiation involves attack at C-4 of the salt and opening of the 1,3-dioxolenium ring with the formation of formic acid derivatives (cf. discussion of the ambident reactivity, p. 21):

$$H-C\underset{O-CH_2}{\overset{O-CH_2}{\diagup\!\!\!\!\diagdown}}{}^+ \Big| \quad , A^- \; + \; O\underset{CH_2-CH_2}{\overset{CH_2-CH_2}{\diagup}} \; \overset{k_i}{\rightleftharpoons} \; \underset{H}{\overset{O}{\diagdown}}C-O-CH_2CH_2-\overset{+}{O}\underset{CH_2-CH_2}{\overset{CH_2-CH_2}{\diagup}} \; A^-$$

a   b   c   d   e   f   g   h   i (60)

Table 6. Rate constants and activation parameters of various initiating systems

Monomer	Initiator	$k_i \cdot 10^3$ (mole^{-1} · l · s^{-1})	Solvent	Temp. (°C)	$\Delta H^{\ddagger}$ (kcal · mole^{-1})	$\Delta S^{\ddagger}$ (cal · mole^{-1} · K^{-1})	Method	Ref.
THF	Et$_3$O$^+$ BF$_4^-$	0.7	CH$_2$Cl$_2$	25	16.4	−16	^{1}H-NMR	87)
THF	MeOSO$_2$CF$_3$	0.4	CH$_2$Cl$_2$	25	6.5	−52	^{1}H-NMR	98)
THF	EtOSO$_2$CF$_3$	0.17	CH$_2$Cl$_2$	25	10.5	−41	^{1}H-NMR	98)
THF	EtOSO$_2$CF$_3$	0.06	CCl$_4$	25	12.5	−36	^{1}H-NMR	125)
THF	EtOSO$_2$CF$_3$	0.04	CCl$_4$	25	20(10)^a	−5(−45)^a	^{19}F-NMR	126)
THF	EtOSO$_2$CF$_3$	0.6	CH$_3$NO$_2$	25	−	−	^{1}H-NMR	98)
THF	EtOSO$_2$O	0.02	CCl$_4$	25	14	−33	^{19}F-NMR	127)
THF	Ph$_3$C$^+$, SbF$_6^-$	3.9 (k$_H$)	CH$_2$Cl$_2$	18	−	−	Polarogr.	60)
THF	Ph$_3$C$^+$, SbF$_6^-$	6.3 (k$_H$)	CH$_2$Cl$_2$	25	12	−26	UV	62)
THF	PF$_5$	0.005	CH$_2$Cl$_2$	25	−	−	^{19}F-NMR	110)
THF	Et$_3$O$^+$BF$_4^-$	0.25	C$_2$H$_4$Cl$_2$	0	−	−	^{14}C	128)
THF	⟨ring⟩ , AsF$_6^-$	0.9	CH$_3$NO$_2$	0	−	−	^{1}H-NMR	129)
OXP	Et$_3$O$^+$BF$_4^-$	0.51	CH$_2$Cl$_2$	25	16.2	−17	^{1}H-NMR	87)
OXP	⟨ring⟩ , AsF$_6^-$	4.5	CH$_3$NO$_2$	25	6.1	−42	^{1}H-NMR	129)
OXP	MeOSO$_2$CF$_3$	2.1	CH$_3$NO$_2$	35	−.	−	^{19}F-NMR	129)
OXIR	EtOSO$_2$CF$_3$	0.41	CH$_2$Cl$_2$	35	−	−	^{19}F-NMR	130)
BCMO	EtOSO$_2$CF$_3$	1.4	C$_6$H$_5$NO$_2$	70	12.5	−36	^{1}H-NMR	131)
1,3-dioxolane	Ph$_3$C$^+$SbF$_6^-$	12 (k$_H$)	CH$_2$Cl$_2$	25	15.6	−15	UV	51)
1,3-dioxolane	PH$_3$C$^+$SbF$_6^-$	7.8 (k$_H$)	CH$_2$Cl$_2$	22	−	−	Polarogr.	60)
1,3-dioxolane	⟨ring⟩ SbF$_6^-$	0.23	CH$_3$NO$_2$	25	−	−	^{1}H-NMR	71)

2-Oxz	MeOTs	0.098	CD$_3$CN	40	—	—	^{1}H-NMR	207)
2-Oxz	MeI	0.16	CD$_3$CN	40	—	—	^{1}H-NMR	207)
2-Me-2-Oxz	MeOTs	0.18	CD$_3$CN	40	13.9	—	^{1}H-NMR	207)
2-Me-2-Oxz	PhCH$_2$Br	0.13	CD$_3$CN	40	—	—	^{1}H-NMR	99)

a   Values recalculated from data of Ref. 126

THF	tetrahydrofuran
OXP	oxepane
OXIR	oxirane
BCMO	3,3-bis-(chloromethyl)oxetane
2-Oxz	2-oxazoline
MeOTs	methyl p-toluenosulfonate

^{1}H-NMR (in ppm δ): a: 9.4 (s), b: 5.6 (s), c: 3.7 (m), d: 1.8 (m), e: 8.1 (s), f: 4.3 (m), g, h: 4.9 (m), i: 2.5 (m). If the rate constant of initiation is lower than that of propagation, an induction period appears on the otherwise linear semilogarithmic plot of monomer consumption ($\ln [([M]_0 - [M]_e)/([M]_t - [M]_e)]$ vs. time). The analysis of these curves, according to the Beste and Hall treatment [124, 125], gives k_i

$$\ln \left\{ \frac{d \ln ([M]_t - [M]_e)}{[I]_0 \, dt} + k_p \right\} = \ln k_p - k_i \int_0^t [M]_t \, dt \tag{61}$$

The initiation rate constant k_i is obtained from the slope of the plot of the left-hand side of Eq. (61) vs. $\int_0^t [M]_t \, dt$. Determined in this way, the value of k_i in the polymerization of THF is in good agreement with k_i found from model studies. The same treatment was applied earlier in the polymerization of 3,3-bis-(chloromethyl)-oxetane [122].

When initiation is more complex, the elementary reactions can sometimes be studied separately. This is the case for initiation of the polymerization of 1,3-dioxolane with trityl salts. In the first reaction, hydride transfer takes place and then the newly formed cation reacts again with monomer. This second process is considerably slower than the former one. The first hydride abstraction was studied by UV [59] (disappearance of the trityl cation absorption at $\lambda = 430$ nm, $\epsilon_{max} = 3.6 \cdot 10^4$) and by polarography [60] (observation of $(C_6H_5)_3C^+$ giving a reversible one-electron wave for the trityl ion reduction with $E_{1/2} \sim 0.51$ V).

There are still a number of relatively simple but unanswered questions arising when the initiation of model monomers is studied. For instance, the rate constants are known neither for the reaction of THF with the secondary oxonium ion H—O⁺⟩

nor for the R—C(=O)—O⁺⟩ cation.

In Table 6 some of the values of the rate constants of initiation determined to date and their corresponding activation parameters are listed.

Results given in Table 6 indicate that there is a good agreement between the data observed by various authors. Thus, for instance, both UV and polarographic methods give very similar results for hydride transfer.

Rate constants of cationation do not markedly differ with varying structure of the nucleophile; indeed, cationation of THF, oxepane and DXL with 1,3-dioxolenium salts with AsF_6^- anions are equal to $0.9 \cdot 10^{-3}$, $1.1 \cdot 10^{-3}$ and $0.23 \cdot 10^{-3}$ (in mole$^{-1} \cdot 1 \cdot s^{-1}$, at $0 \,^\circ$C in CH_3NO_2), respectively. The closeness of these values indicates, that the bond breaking in the cation is much more advanced in the transition state than the bond making.

Rate constants of cationation with esters of triflic acid depend on the polarity of the medium (e.g. for THF at 25 $^\circ$C $k_i = 1.7 \cdot 10^{-4}$ mole$^{-1} \cdot 1 \cdot s^{-1}$ in CH_2Cl_2 but only $0.6 \cdot 10^{-4}$ mole$^{-1} \cdot 1 \cdot s^{-1}$ in CCl_4) and also slightly on the monomer used.

Unfortunately, there are no data for triflic anhydride $(CF_3SO_2)_2O$ available which, according to the existing reports, gives rise to much faster initiation than esters.

In our opinion, the anhydride when used in CH_2Cl_2 or CH_3NO_2 solvents, should replace BF_3 and its complexes for screening experiments of new monomers. At least in one report [132], in which the anhydride was directly compared with BF_3 complexes and trityl salts (copolymerization of DXL with 1,3,5-trioxane) the anhydride was found to be far superior giving higher polymerization degrees and much higher rates of polymerization than BF_3 complexes. It was possible in this system to use 10^2 times lower concentration of the triflic anhydride than that of BF_3 $(n\text{-}C_4H_9)_2O$ initiator and obtain comparable overall rates of polymerization.

4 Propagation

Despite the fast progress of our understanding of the cationic polymerization of heterocycles, the number of systems providing truly living polymers are still very limited. However, the proper choice of the necessary conditions for a given monomer (solvent, temperature, anion) gives an access to studies of propagation reactions in optimized systems, i. e. with propagation accompanied by another reaction (e. g. initiation or termination) that could also be studied quantitatively.

4.1 Chemical Structure of the Growing Species

Structures of the growing species, assumed in earlier works, have been directly observed by NMR in more recent studies. Taking the polymerization of THF as an example, the growing tertiary oxonium ion as an active species was postulated as one of the possible structures first by Meerwein[77] and than by other groups describing the cationic polymerization of THF in terms of a living process[128, 133–137].

NMR studies of the polymerization of THF, oxepane, 3,3-dimethylthietane, 2-methoxy-2-oxo-1,3,2-dioxaphosphorinane and 2-methyl-2-oxazoline revealed that the structures of the growing species are the corresponding onium ions; these structures are given in Table 7 (earlier, the first reaction products, formed upon initiation, were observed in the polymerization of THF[78]).

In Sect. 3.2, the carbenium-onium equilibria were discussed; onium ions listed in Table 7 can also be components of this equilibrium and it may be argued that the actual propagation involves a minute concentration of the much more reactive carbenium ions, being in equilibrium with an overwhelming concentration of the less reactive onium ions. Thus, in e. g. the polymerization of THF, the two following rival chain propagations have to be considered first of all (the same discussion could apply to any other growing onium species):

$$
\text{+} \cdots \text{)}_{\overline{n}}\text{O}-\text{CH}_2-\text{CH}_2-\text{CH}_2-\overset{+}{\text{CH}}_2 \ + \ \text{O}\bigcirc \ \underset{\text{slow}}{\overset{\text{fast}}{\rightleftharpoons}} \ \text{+} \cdots \text{)}_{\overline{n}}\text{O}-\text{CH}_2-\text{CH}_2-\text{CH}_2-\text{CH}_2-\overset{+}{\text{O}}\bigcirc
$$

$$
\text{+} \cdots \text{)}_{\overline{n}}\text{O}-\text{CH}_2-\text{CH}_2-\text{CH}_2-\text{CH}_2-\overset{+}{\text{O}}\bigcirc \ \underset{\text{fast}}{\overset{\text{slow}}{\rightleftharpoons}} \ \text{+} \cdots \text{)}_{\overline{n+1}}\text{O}-\text{CH}_2-\text{CH}_2-\text{CH}_2-\overset{+}{\text{CH}}_2
$$

(62)

This is an analog of the S_N1 reaction (of A_1 in acid hydrolysis). Thus, the rate of the overall reaction would be controlled by a slow unimolecular ring-opening within an oxonium ion which, in turn, is formed in the fast reaction with the next monomer molecule.

Table 7. Growing species in cationic polymerization of heterocyclic monomers observed directly by ^{1}H-NMR spectroscopy[a]

Monomer	Structure of growing species (anions omitted)	Chemical shifts δ (ppm) from TMS	Ref.
 (tetrahydrofuran ring)	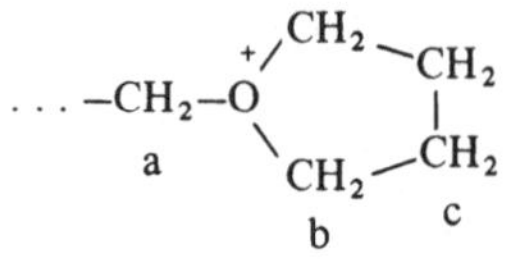 $\cdots -CH_2-O^+$ a, ring CH_2-CH_2 (b), CH_2-CH_2 (c)	a 4.89 (t)[b] b 4.86 (t) c 2.41 (q)	138)
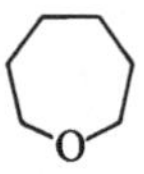 (oxepane ring)	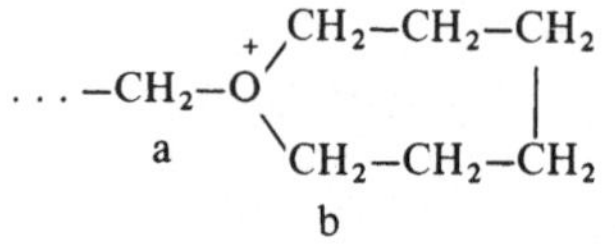 $\cdots -CH_2-O^+$ a, $CH_2-CH_2-CH_2$ / $CH_2-CH_2-CH_2$ (b)	a 4.97 (t) b 5.10 (t)	129)
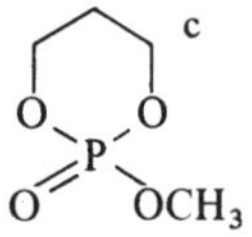 (dioxaphosphorinane, O,O–P(=O)OCH₃)	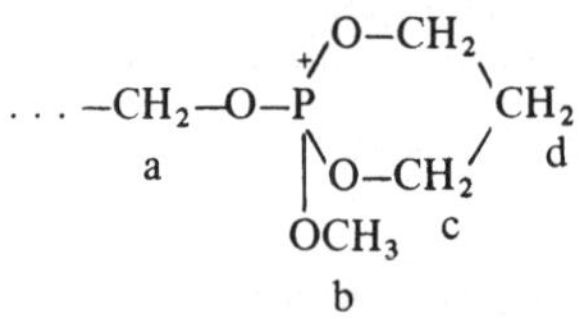 $\cdots -CH_2-O-P^+$ a; $O-CH_2$ / $O-CH_2$ (b); CH_2 (d); OCH_3 (c)	a 4.7 (m)[c] b 4.2 (s)[c] c 5.05 (t)[dc] d 2.7 (quint)[c]	139)
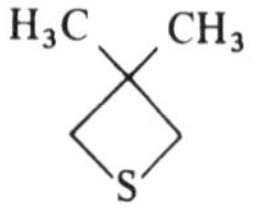 H₃C, CH₃ (thietane, S)	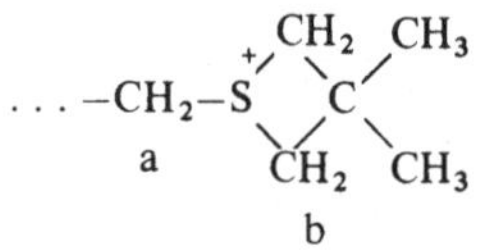 $\cdots -CH_2-S^+$ a; $C(CH_2)(CH_3)$; CH_2 CH_3 (b)	a 3.80 (s) b 3.66, 3.94 (AB system)	140)
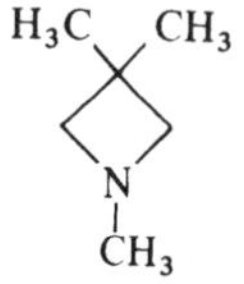 H₃C, CH₃ (azetidine, N–CH₃)	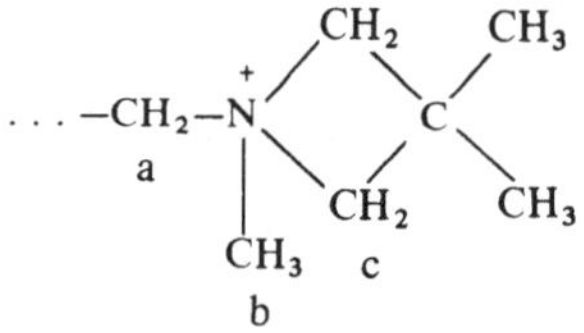 $\cdots -CH_2-N^+$ a; CH_2 CH_3; C; CH_2 CH_3; CH_3 (b), CH_2 (c)	a 3.88 (s) b 3.74 (s) c 4.55, 4.67 (AB system)	141)
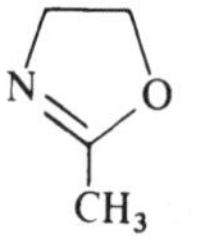 (2-methyl-oxazoline, N=C–CH₃, O)	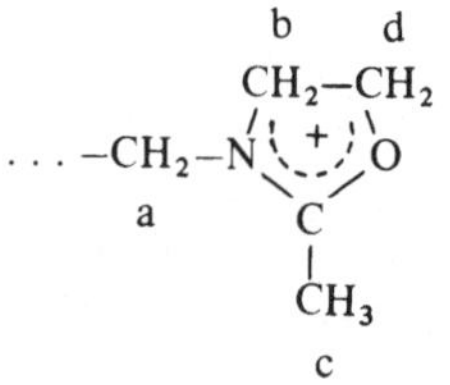 $\cdots -CH_2-N^+$ a; CH_2-CH_2 (b,d); $C-CH_3$ (c); O	a ~3.5 (m)[e] b 4.2 (t)[e] c 2.55 (s)[e] d ~4.8 (t)[e]	99)

[a] Chemical shifts can slightly depend on concentration and solvent used

[b] Difference between the exocyclic (a) and endocyclic (b) protons could only be observed at high resolution

[c] ^{31}P decoupled

[d] The presence of the undistorted triplet and quintuplet indicates that the tetraalkoxyphosphonium cation is conformationally labile

[e] In spite of further isomerization in the propagation step, no isomerized growing species like

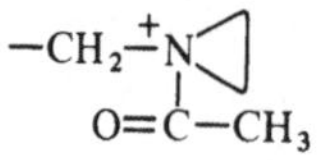

$$-CH_2-\overset{+}{N}\diagup \quad \text{(with } O=\overset{|}{C}-CH_3 \text{)}$$

have been observed

The second mechanism, which seems to be generally accepted, refers to its counterpart in nucleophilic substitution, namely to the S_N2 reaction:

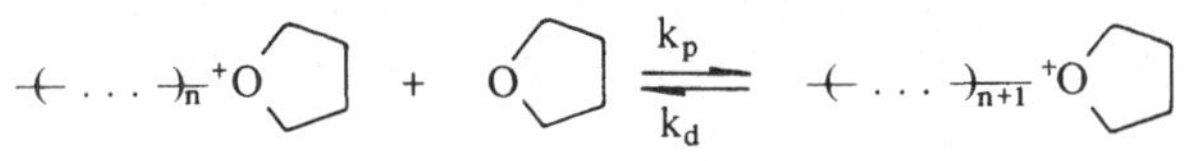

$$\longleftarrow \cdots \rightarrow_{\overline{n}}^{+}O\bigcirc \quad + \quad O\bigcirc \quad \underset{k_d}{\overset{k_p}{\rightleftharpoons}} \quad \longleftarrow \cdots \rightarrow_{\overline{n+1}}^{+}O\bigcirc \tag{63}$$

The stereochemical course of this reaction requires a linear transition state; this nucleophilic attack, as shown below, leads to the inversion of configuration at the carbon atom in the α-position of the oxonium ion:

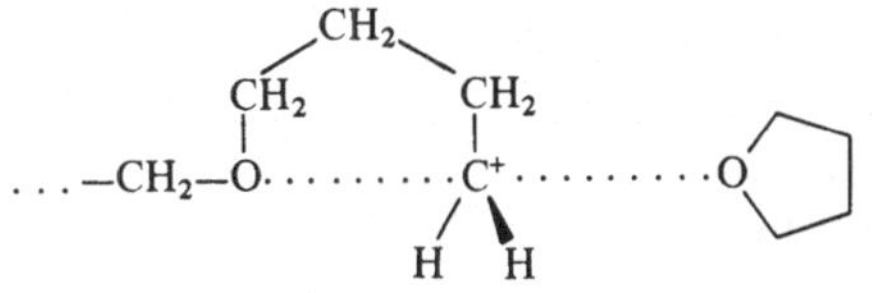

$$\cdots -CH_2-O\cdots\cdots\cdots\overset{\overset{\displaystyle CH_2}{\overset{\displaystyle |}{\underset{\displaystyle \underset{H\quad H}{\diagup\diagdown}}{C^+}}}}{}\cdots\cdots\cdots O\bigcirc \tag{64}$$

A number of arguments support this mechanism for the polymerization of cyclic ethers:

1. Carbenium ions that should operate in the S_N1-like mechanism are strong hydride abstractors; thus, the primary carbenium ions easily rearrange through the intramolecular hydride shift to the more stabilized secondary or tertiary carbenium ions (reactions observed in e. g. cationic vinyl polymerization)[142]. Neither hydride abstraction including intermolecular transfer nor isomerization (the latter argument has already been used by Medvedev a. o.[143]) have been observed in the polymerization of THF, even in systems kept at the living polymer-monomer equilibrium for a longer time.

2. Equilibria like (62) are strongly shifted to the side of oxonium ions; e. g. the calculated heat of formation of the triethyloxonium ion from ethyl cation and diethyl ether is as high as 128 kcal $\cdot$ mole^{-1} in the gas phase[144]. This would give no chance for a single carbenium ion to exist at the usual polymerization conditions of the majority of heterocycles: cyclic sulfides and amines are even stronger nucleophiles than cyclic ethers (cf., however, the discussion on cyclic acetals below).

3. In cationic polymerization, tert-butyloxirane opens, according to Tsuruta[145],

exclusively at the $-CH_2\overset{\downarrow}{-}O$-bond and not at the $(H_3C)_3C\overset{|}{-}\overset{\downarrow}{CH}\overset{|}{-}O$-bond as it could be assumed for the S_N1 mechanism.

4. The polymerization of cyclic ethers like oxirane[146], substituted oxiranes[147] and bicyclic tetrahydrofurans[148] proceeds with inversion of configuration on the carbon atom, strongly suggesting the S_N2 mechanism.

4.1.1 Active Species in the Polymerization of Cyclic Acetals

It has been shown in the previous sections that at least the polymerization of cyclic ethers, sulfides and amines proceeds via onium ions. The large majority of authors have agreed on this point (the mechanism of propagation of disubstituted cyclic ethers like isobutylene oxide is, however, still in dispute)[147].

In the polymerization of cyclic acetals there has been a long lasting discussion about the nature of growing species and it seems to be at present only partially resolved.

1,3-Dioxolane was used as the model monomer and there are two groups of opinions about the structure of the growing species in polymerization. One group of authors maintains that poly-1,3-dioxolane grows exclusively as a macro-ring[149—151] with insertion of the monomer molecules in the propagation step and with release of the monomer molecule from the growing macro-ring in the depropagation step. This theory can schematically be represented as follows (for 1,3-dioxolane):

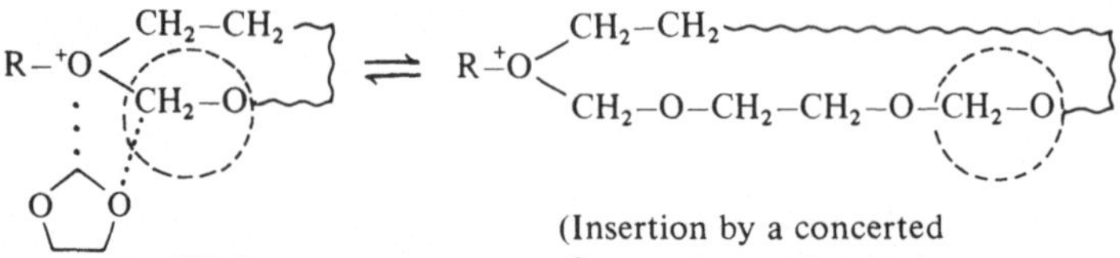

(65)

(Insertion by a concerted
four-center mechanism)

Another group[51, 152—156] sees propagation of 1,3-dioxolane in more conventional terms assuming that not cyclic but linear macromolecules grow, having active centers alkoxycarbenium or oxonium at their ends:

alkoxycarbenium structure   oxonium structure

IV.2   IV.3

For a moment, we make only this distinction, namely between the ring-expansion and the linear growth mechanisms; later, the carbenium-oxonium controversy will be treated.

First of all, it will be shown that the recent direct NMR observation of the growing species has allowed a discrimination between these two concepts; indeed, propagation involves linear macromolecules. Then, we shall proceed with the discussion of the more exact structure of the growing species in these linear macromolecules.

With each passing day and with new data the involved authors took up one position or another; thus, we shall refrain from quoting all the arguments used in the past, especially since in the light of the direct NMR observations, there is no reason to call for the more speculative explanations.

4.1.2 Polymerization of Cyclic Acetals Initiated by Protonic Acids

Macrocyclic structure IV.1, supposedly growing by ring-expansion, requires that a proton would exclusively constitute a component of the secondary oxonium ion both during propagation and after establishment of the polymer-monomer equilibrium:

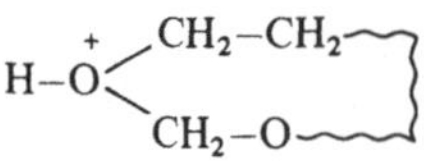

IV. 1

If, however, linear growing macromolecules are present, then protons would constitute also or exclusively the primary hydroxy end groups: $HO-CH_2-CH_2-O-CH_2$ e. g.:

$$
\begin{array}{ccc}
\text{IV. 1} & \rightleftharpoons & \text{IV. 4}
\end{array}
\qquad (66)
$$

Chemical shifts of protons in secondary oxonium ions differ substantially from chemical shifts of protons in the primary hydroxy groups. One can expect a fast proton exchange between these two species. However, if the individual chemical shifts are known, then the observed chemical shift (due to exchange) permits the determination of the actual proportions of the secondary oxonium ion $IV.1$ and tertiary oxonium ion $IV.4$.

The chemical shift of the protons under discussion as a function of the ratio [1,3-dioxolane]$_0$/["H" from initiator]$_0$ is shown in Fig. 1 taken from Ref. 34 (CF_3SO_3H was used as the acid).

From the ratio [1,3-dioxolane]$_0$/[$CF_3SO_3H_0$] $\simeq 1.0$ up to ~ 5, the chemical shift is equal to $\delta = 14.5$ ppm, being close to the value known for the secondary oxonium ion:

$$
CF_3-SO_3H + \overset{\frown}{O \quad O} \;\overset{K_1}{\rightleftharpoons}\; CF_3-SO_3H \ldots \overset{\frown}{O \quad O} \;\overset{K_2}{\rightleftharpoons}\; \underbrace{H-\overset{+}{O} \overset{\frown}{\quad} O \; CF_3-SO_3^-}_{\delta = 14.5 \text{ ppm}} \qquad (67)
$$

When the concentration of [1,3-dioxolane]$_0$ reaches its equilibrium concentration, polymerization ensues and the chemical shift moves upfield, i. e. in the direction of the chemical shift of the hydroxy group (structure $IV.4$). This indicates that the polymerizing mixture contains tertiary oxonium ions and that their proportion depends on the average length of the macromolecule. This is so because the longer the chain, the lower the probability of end-biting, producing secondary oxonium ions. (For a more detailed discussion of the dependence of end-biting on chain length see Ref. 34.)

The induction periods observed in the polymerization of 1,3-dioxolane initiated by acids (e. g. perchloric acid[149], trifluoromethanesulfonic acid[34]) is apparently due to the equilibrium shifted toward much less reactive secondary oxonium ions at the early stages (short chains) of polymerization.

The simultaneous presence of secondary and tertiary oxonium ions in the polymerization of 1,3-dioxolane initiated with trifluoromethanesulfonic acid was also shown by trapping a proton (from the secondary oxonium ion) and a carbenium ion (from either the tertiary oxonium ion or the alkoxycarbenium ion) by the ion-trapping technique.

Let us consider again the equilibrium involved between the rival species discussed[156]:

$$
\begin{array}{c}
\text{H-O} \quad \text{O} \quad \overset{+}{\text{O}} \quad \text{O}\diagdown \\
\text{CH}_2\text{-O}\diagdown
\end{array}
$$

IV. 4

$$
\begin{array}{ccc}
\text{H-}\overset{+}{\text{O}} \quad \text{O} \quad \text{O} \quad \text{O}\diagdown & & \text{H-O} \quad \text{O} \quad \text{O} \quad \text{O}\diagdown\diagdown\text{O-CH}_2\text{-}\overset{+}{\text{O}}\diagup \\
\text{CH}_2\text{-O}\diagdown\diagdown & \rightleftharpoons & \\
\textit{IV. 1} & & \textit{IV. 5}
\end{array}
$$
\hfill (68)

(where $\diagup\!\!\!\!\text{O}\diagdown$ denotes a part of the monomer molecule or of the foreign macromolecule).

Ion-trapping[157] is based on the fast reactions between cations and trialkylphosphines leading to stable phosphonium salts, namely tertiary phosphonium salts when a proton is trapped (e. g. from *IV.1*) and quaternary salts when the reaction involves tertiary oxonium ions (e. g. *IV.4* and/or *IV.5*) or alkoxycarbenium ions.

^{31}P-NMR spectra of the polymerization mixture consisting of 1,3-dioxolane initiated with CF_3SO_3H and trapped with $(n\text{-}C_4H_9)_3P$ reveal the presence of both $H\text{-}\overset{+}{P}(n\text{-}C_4H_9)_3$ and $\sim\sim O\text{-}CH_2\text{-}\overset{+}{P}(n\text{-}C_4H_9)_3$ species ($\delta = -11.9$ ppm and -31.4 ppm, respectively from H_3PO_4 in CH_2Cl_2 solvent at $-70\,°C$)[158]. The proportions of the tertiary phosphonium cations and of the quaternary cations depend on the initial ratio of monomer to initiator. For short chains (i. e. at low $[\text{monomer}]_0/[\text{initiator}]_0$ ratio), tertiary phosphonium cations prevail indicating that secondary oxonium ions dominate in the system. Secondary oxonium ions are either cyclic (like *IV.1*) or linear; this latter structure is observed when a proton is a part of secondary oxonium ion along linear macromolecule.

$$
\text{H-O} \quad \text{O} \quad \text{O} \quad \overset{+}{\underset{\text{H}}{\text{O}}}\sim\!\sim\!\sim\sim\text{O-CH}_2\text{-}\overset{+}{\text{O}}\diagup
$$

Thus, the proportion of secondary oxonium ions, found by ^{1}H- and ^{31}P-NMR, is equal to the sum of those being part of a macrocycle and those belonging to linear macromolecules.

Propagation may proceed by both secondary and tertiary oxonium ions but the latter are known to be much more reactive. The respective rate constants have not yet been determined.

4.1.3 Polymerization of Cyclic Acetals Initiated by Preformed or "Hidden" Carbocations

In the polymerization of 1,3-dioxolane initiated with triethyloxonium salt it has been shown by ^{1}H-NMR that the ethyl groups stemming from the initiator are bound exclusively as uncharged ethoxy end-groups when initiation has been completed. No ethyl group bonded to an oxonium ion could be observed, although the initiating triethyloxonium ion was clearly detected at the begining of polymerization[156]. Deuterated 1,3-dioxolane was used to decrease absorptions coming from backbone and equilibrated monomer:

$(CH_3CH_2)_3O^+A^-$ $CH_3CH_2OCD_2CD_2OCD_2\sim\!\!\sim\!\!\sim\!\!\sim\!\!\sim\!\!\sim$ ⊕ A^-

δ (ppm): 1.75 (t)   1.17 (t)   (A = SbF_6)

In the polymerization conducted with $3 \cdot 10^{-2}$ mole $\cdot$ 1^{-1} of $(C_2H_5)_3O^+SbF_6^-$ and 4.0 mole $\cdot$ 1^{-1} of 1,3-dioxolane $-d_6$, initially only a triplet was observed at δ = 1.75 ppm. This was absent in the living equilibrated system and only a triplet at δ = 1.17 was found (together with a triplet from the liberated diethyl ether centered at δ = 1.14 ppm)[156]. The chemical shift of the protons of the ethyl group located on the oxonium ion in the macro-ring should be close to that in the triethyloxonium salt. The absence of absorption in this region in the equilibrated system indicates that the growing species are predominantly linear.

4.1.4 Carbenium or Oxonium Ions in the Polymerization of Cyclic Acetals?

In the preceding sections, NMR data were discussed indicating that poly-1,3-dioxolane grows on the linear macromolecules. The corresponding active species can either be oxonium ions or alkoxycarbenium ions:

$$\ldots -O-CH_2-CH_2-O-CH_2-\overset{+}{O}\!\!< \; \rightleftharpoons \; \ldots -O-CH_2-CH_2-O-\overset{+}{C}H_2 + O\!\!< \qquad (69)$$

(where $O\!\!<$ is either a part of its own (i. e. cyclic structure) or foreign macromolecule or monomer molecule).

Again, a number of indirect observations were used for and against both of the structures in Eq. (69) but there is no clear-cut evidence for any system suggesting that exclusively one of these two species are present.

4.1.5 Carbenium-Oxonium Ion Equilibria Involving Alkoxycarbenium Ions

Recently, it was shown by ^{1}H-NMR on the basis of direct studies of the $CH_3O\overset{+}{C}H_2$ cation (being a model of the supposed growing alkoxycarbenium species) that its exchange with the corresponding oxonium counterpart was very fast[159].

For the system

$$CH_3-O-\overset{+}{C}H_2 \; + \; O\!\!<\!\!\begin{array}{c}CH_3\\CH_2\end{array} \quad \underset{k_d}{\overset{k_a}{\rightleftharpoons}} \quad CH_3-O-CH_2-\overset{+}{O}\!\!<\!\!\begin{array}{c}CH_3\\CH_2\end{array} \qquad (70)$$

$$O\!\!<\!\!CH_3 O\!\!<\!\!CH_3$$

$$IV. 6 IV. 7$$

and

$$CH_3-O-CH_2-\overset{+}{O}\!\!<\!\!\begin{array}{c}CH_3\\CH_2\\O\!\!<\!\!CH_3\end{array} \; + \; O\!\!<\!\!\begin{array}{c}CH_3\\CH_2\\O\!\!<\!\!CH_3\end{array} \quad \underset{k_{ex}}{\overset{k_{ex}}{\rightleftharpoons}} \quad CH_3-O-CH_2-\overset{+}{O}\!\!<\!\!\begin{array}{c}CH_3\\CH_2\\O\!\!<\!\!CH_3\end{array} \; + \; O\!\!<\!\!\begin{array}{c}CH_3\\CH_2\\O\!\!<\!\!CH_3\end{array}$$

IV. 7   (71)

it was found: $k_a = 2 \cdot 10^6 \text{ mole}^{-1} \cdot 1 \cdot s^{-1}$, $k_d = 7 \cdot 10^2 \cdot s^{-1}$,

$k_{ex} = 1.9 \cdot 10^4 \text{ mole}^{-1} \cdot 1 \cdot s^{-1}$,

$$K_e = k_a/k_d = \frac{[\text{oxonium}]}{[\text{alkoxycarbenium}] \cdot [\text{dimethoxymethane}]} = 3 \cdot 10^3 \text{ mole}^{-1} \cdot 1$$

The above data were found in SO_2 solution at $-70\,°C$ from the studies of line-broadening and averaged chemical shifts[159].

It is remarkable that the rate constant k_a of the addition of dimethoxymethane to methoxycarbenium ion *IV.6* is only 10^2 times higher than the rate constant k_{ex} of the identical reaction involving the equivalent oxonium ion *IV.7*.

Under typical polymerization conditions, the total concentration of growing species is e.g. 10^{-3} mole $\cdot 1^{-1}$ and the concentration of polymer is equal to e.g. 2.5 mole $\cdot 1^{-1}$. Under these conditions, taking into account the value of $K_e = 3 \cdot 10^3$ mole^{-1}, we would have 10^{-3} mole $\cdot 1^{-1}$ of oxonium ions (assuming that dimethoxymethane is a suitable model for the polymer, and that polymer and monomer have identical K_e values) and only $1 \cdot 10^{-7}$ mole $\cdot 1^{-1}$ of polymeric alkoxycarbenium ions. Thus, for every 1000 monomer molecules converted into polymer, only 10 would be added through carbenium ions and the rest through oxonium ions. The actual data for 1,3-dioxolane are not known at present and may differ from values found for dimethoxymethane. Nevertheless, of primary importance is the finding that the reactivities of alkoxycarbenium ions and tertiary oxonium ions toward linear acetals, expressed through the corresponding rate constants, differ for the discussed above conditions only by 10^2 times.

The proportions of oxonium and carbenium ions in the chain growth, as well as their relative reactivities, may markedly depend on the polymerization conditions and vary with monomer structure. The high values of rate constants of the reaction between carbenium cations and ethers or acetals make questionable Okada's interpretation of the 1,3-dioxane-triethyloxonium salts system[159]. It is impossible to observe, at room temperature, the individual $ROCH_2^+$ species with excess of such nucleophiles as diethyl ether. We would rather suggest that the signal at $11-13$ ppm $\delta(^1\text{H-NMR})$ or $175-180$ ppm $\delta(^{13}\text{C-NMR})$ is due to some exchanging protonated species.

4.1.6 Structure of Tertiary Oxonium Ions in the Polymerization of 1,3-Dioxolane

It has already been discussed in the preceding sections that, at least in the case of 1,3-dioxolane, the growing oxonium ions are involved in multiple equilibria, including the larger cyclic oxonium ions (e. g. 7-membered ones).

Thus, we propose to describe the active species in the polymerization of 1,3-dioxolane as a tertiary oxonium ion:

$$\ldots -O-CH_2-O-CH_2-CH_2-O-CH_2-\overset{+}{O}\!\!<$$

in which the $O\!\!<$ moiety designates either a part of its own or of a foreign macromolecule. This structure is similar to that proposed earlier[51], since it has to be remembered that the alkoxycarbenium residue of this structure can change its position along a macromolecule several times during the time elapsing between two successive monomer additions.

Thus, one of the possible propagation steps can be visualized as follows:

$$\ldots -O-CH_2-O-CH_2-CH_2-O-CH_2-\overset{+}{O}\diagdown \ + \ \diagup O \qquad O \diagdown \longrightarrow$$

$$\longrightarrow \ldots O-CH_2-O-CH_2-CH_2-O-CH_2-{}^+O \diagup \qquad O \diagdown \longrightarrow \qquad (72)$$

$$\xrightarrow{\text{fast isomerization}} \ldots -O-CH_2-O-CH_2-CH_2-{}^+O \qquad \xrightarrow[\text{slow propagation}]{}$$

Scheme (72) shows only one of the possible routes accounting for the fact that 7-membered rings are directly observed. Nevertheless, the second step may involve any of the oxygen atoms in the own chain. It is not known at present how large the preferred ring is and only the equilibrium between 5- and 7-membered rings has been measured. The high rate of the second step (compared with similar but bimolecular reactions with monomer) is ensured by the anchimeric assistance (neighboring group participation) and higher nucleophilicity of linear acetals. This second reason may also lead to the presence of the tertiary oxonium ions in which two different macromolecules are connected to one another via the tertiary oxonium ion as a junction point. The proportion of these various oxonium ions in propagation may differ for various monomers and even for a given monomer, depending on polymerization conditions.

Recently, kinetic isotope effects were studied in the polymerization of DXL. The results reported indicate that the ring opening of the monomer molecule does not occur in the rate-limiting step of the propagation process and thus support the growth on the cyclic tertiary oxonium ion[160].

This lengthy discussion of the structure of growing species in the polymerization of cyclic acetals clearly indicates that this is, to some extent, still an open question although two of the hypotheses, namely growth by ring expansion and growth on carbenium ions are almost eliminated (at least for DXL). Thus, one can conclude that it is fairly well established that polymerization of DXL proceeds on linear growing species having the structure of tertiary oxonium ions. A small proportion of the polymeric alkoxy carbenium ions may coexist in equilibrium with the tertiary oxonium ions, but their contribution in the chain growth is not important, at least at lower temperatures.

Polymerization initiated by protonic acids additionally involves an equilibrium between secondary and tertiary oxonium ions, with proportions of both kinds depending on the chain length. For the longer chains, the proportion of the tertiary oxonium ions increases.

If, during initiation, another highly nucleophilic end group is introduced, which is much more nucleophilic than oxygen atoms in the backbone, then end-biting can dominate over back-biting involving any of the monomer units along the macromolecule.

The main differences between oxonium ions in the polymerization of THF and oxonium ions in the polymerization of DXL can be summarized as follows:

— in the polymerization of THF the growing oxonium ion has a uniform structure during the build-up of the given macromolecule; one tertiary oxonium ion is converted in the propagation step into another one of the same chemical structure:

$$\dots -CH_2-CH_2-{}^+O\!\!\bigcirc \;+\; O\!\!\bigcirc \;\rightleftharpoons\; \dots -CH_2-CH_2-O-(CH_2)_4-{}^+O\!\!\bigcirc \qquad (73)$$

— in the polymerization of DXL the oxonium ion has a much more labile structure and between two subsequent monomer additions the alkoxycarbenium moiety of the oxonium ion can change the site of cationation many times, travelling, for instance, along its own macromolecule and changing in this way the attached ligands and forming two other branches of the tertiary oxonium ion.

It cannot be excluded that in the polymerization of cyclic acetals which are less nucleophilic than 1,3-dioxolane, propagation on alkoxycarbenium ions can also have its share in building up the macromolecules (the polymerization of 1,3,5-trioxane, the least nucleophilic acetal, may proceed this way).

4.1.7 Activated Monomer as Active Species in the Polymerization of Lactams

In the cationic polymerization of lactams, initiated with protonic acids, macromolecules are formed having two ends (A and B in the structure below) potentially active[31, 161] (cf. Sect. 3.1):

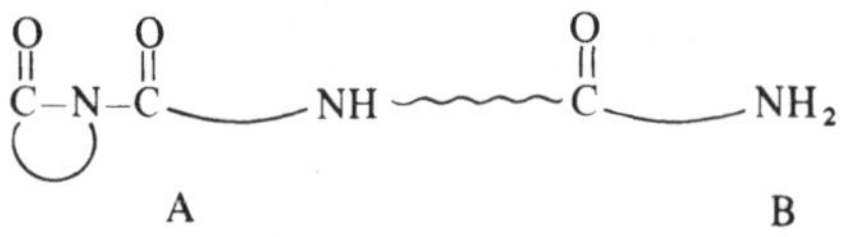

Propagation on both ends A and B involves reaction with an "activated monomer"

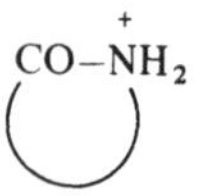

which represents a protonated lactam molecule in the form of the amidium ion. The amidium ion can be in equilibrium with the corresponding hydrogen bonded complex with protonic acid (see however pp. 11 and 134).

The amidium ion is formed directly in the initiation process; then it reacts with monomer giving the first addition product

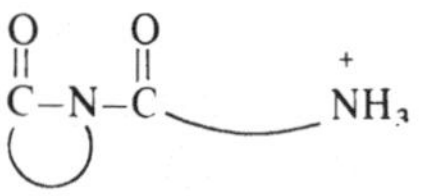

and is regenerated by proton transfer to the next monomer molecule which in this way becomes "activated":

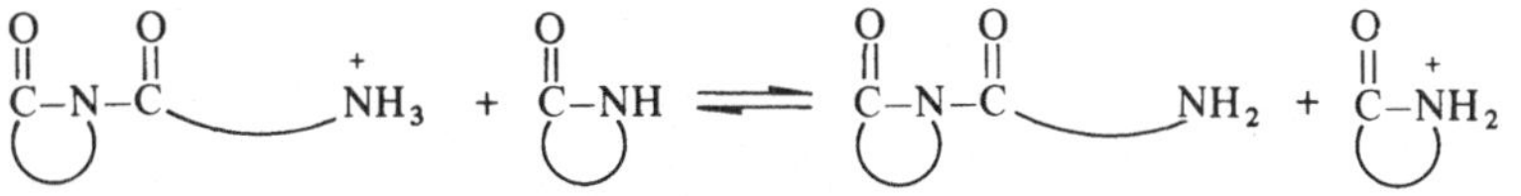

$$\overset{O}{\underset{\underset{\bigcirc}{\parallel}}{C}}-N-\overset{O}{\underset{\parallel}{C}}\underline{\quad\quad}NH_3^+ \ + \ \overset{O}{\underset{\underset{\bigcirc}{\parallel}}{C}}-NH \ \rightleftharpoons \ \overset{O}{\underset{\underset{\bigcirc}{\parallel}}{C}}-N-\overset{O}{\underset{\parallel}{C}}\underline{\quad\quad}NH_2 \ + \ \overset{O}{\underset{\underset{\bigcirc}{\parallel}}{C}}-NH_2^+ \qquad (74)$$

This process of activation can proceed many times at any stage of propagation, as the reversible proton transfer.

The activated monomer may react with both chain ends A and B. Reaction with the chain end B can be presented as follows:

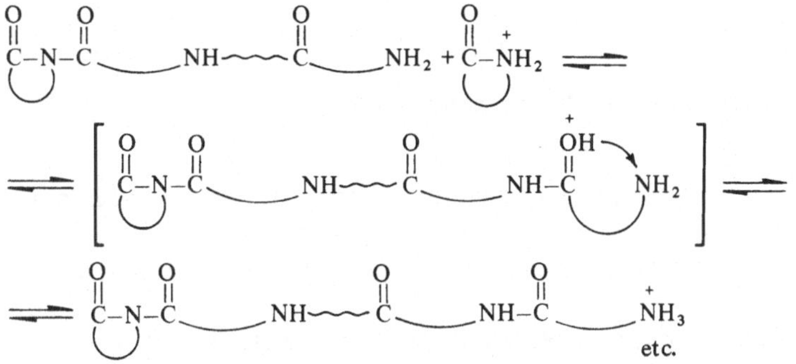

$$(75)$$

This sequence of proton transfer to monomer (formation of activated monomer) followed by attack at the ~NH_2 group from the chain end by the activated monomer and leading to the monomer addition with generation of the ~$\overset{+}{N}H_3$ group thus represents the chain growth on the chain end B.

The chain end A can participate not only in the typical propagation with monomer molecule but it can also be involved in condensation with chain end B:

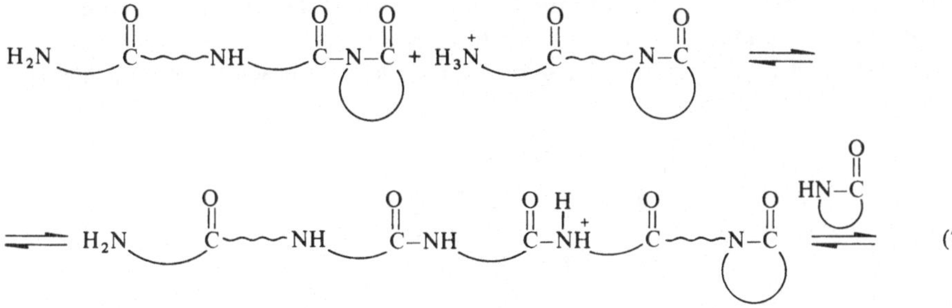

$$(76)$$

Fast proton transfer to monomer regenerates the activated monomer molecule which, in turn, is capable of reacting with a further monomer molecule and of regenerating the consumed reactive chain ends $\sim\sim\overset{+}{N}H_3$ and $\underset{\diagup\diagdown}{\overset{\displaystyle O \atop \parallel}{C-N}}\sim\sim$

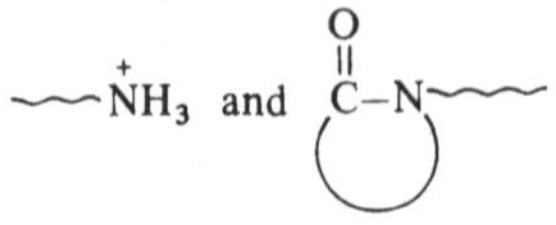

according to reaction (12) in Sect. 3.1.

Taking into account the equilibrium nature of lactam polymerization, there will be always enough monomer available for protonation. However, the proportion of protonated monomer will decrease with conversion.

The chemistry of the above described reactions was verified on the basis of model studies[31, 161].

4.1.8 The Steric and Electronic Structure of Onium Ions

Onium ions, as has been shown in previous sections, are either the exclusive or dominating structures of the growing species in the cationic polymerization of heterocyclic monomers. Literature concerning their steric and electronic structure is, however, rather scarce.

The formation of onium ions requires that one of the lone electron pairs of N, O, P and S atoms in compounds containing C−X−C unit (where X is a heteroatom) is used in the formation of a new bond. In the simplest case, the lone electron pair of the oxygen atom in a water molecule accepts a proton to form a hydronium ion.

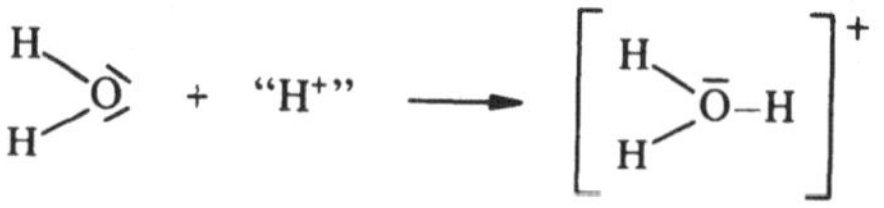

$$\overset{H}{\underset{H}{>}}\ddot{O} \ + \ \text{``}H^+\text{''} \ \longrightarrow \ \left[\overset{H}{\underset{H}{>}}\bar{O}-H\right]^+ \tag{77}$$

The heat of formation of H_3O^+ from the components is 180 kcal · mole^{-1} (the solvation of hydronium ions is discussed in Sect. 3.1).

The electronic change in the formation of the oxonium ions can be visualized in the following way:

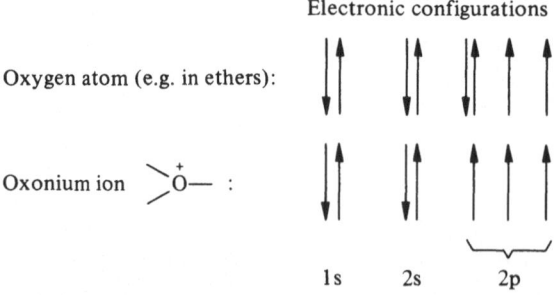

Electronic configurations

Oxygen atom (e.g. in ethers):

Oxonium ion $>\overset{+}{O}-$:

1s   2s   2p

Thus, the bonding in oxonium ions is similar to that in ammonia or trialkylamines having a structure of a trigonal pyramid. The presence of $2p_x$, $2p_y$ and $2p_z$ electrons suggests a trigonal pyramid with bond angles equal to 90°. The actual angles depend, however, on steric factors and increase with increasing repulsion between substituents at the oxygen atom.

Ammonium ions are also formally formed by removal of one electron from the nitrogen atom:

Electronic configurations

Nitrogen atom (e.g. in trialkylamines):

Ammonium cation $\overset{+}{\underset{\cdot}{\text{N}}}$:

2s   2p

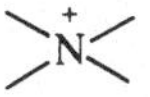

The electronic structure of the ammonium ion is similar to that in the tetrahedral carbon atom and, therefore, sp^3 hybridization becomes possible.

The bond strength in onium ions is even higher than in the starting uncharged compounds[162]. For instance, the strength of the O–H bond is as high as 133 kcal in the hydronium ion while in water it is equal only to 110 kcal. In the ammonium ion NH_4^+ every N–H bond has an energy higher by 66.5 kcal than in NH_3.

The configuration of onium ions stems from their electronic structure. Alkyl groups in oxonium ions are directed along the edges of a pyramid with angles at the top depending, however, in some systems on the steric requirements.

The only oxonium ion studied in detail is cationated oxirane for which the pyramidal structure was proven by NMR studies[163].

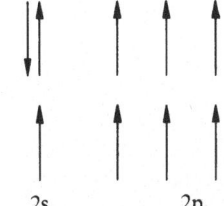

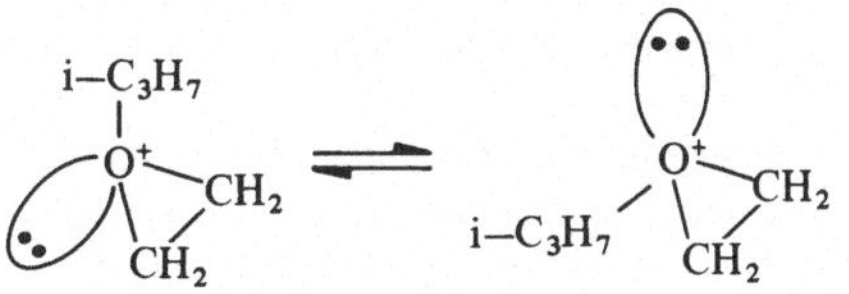

The activation energy for the inversion was found to be 10 ± 2 kcal · mole^{-1}. In SO$_2$ solution at -70 °C, inversion is virtually stopped and the AA'BB' pattern ($J_{AB} \sim 3$ Hz) for ring protons is observed. At 40 °C, ring protons give a sharp singlet indicating that at this temperature inversion is fast relative to the NMR time scale[163].

For tetrahydropyranium ion, inversion is fast even at -70 °C. It is not clear whether these data refer to free or paired cations (SO$_2$ solutions, concentration range 0.2–2 mole · 1^{-1}). One would expect that the conformational lability should be different for the two ionic forms. In the case of an ion pair, one conformation, namely the one in which the anion (counterion) is located on the opposite side to the electron pair, should be favored over the second conformer in which electrostatic repulsion between the electron pair and the negative counterion can be expected. Thus, it would be interesting to study the conformational lability of oxonium ions as a function of the extent of ion pairing.

The exact geometry of oxonium salts is not known. Calculations using the CNDO-2 method indicate the following structures for protonated and methylated oxirane:

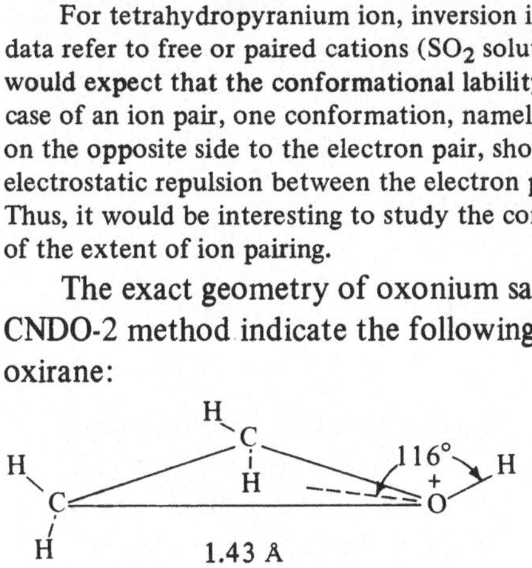

Ref. 164 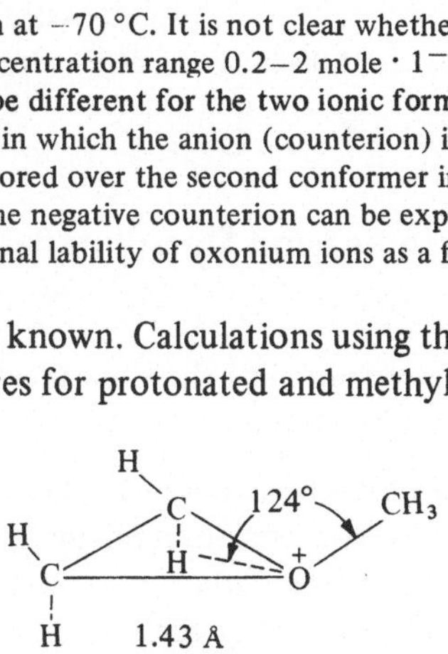   Ref. 165

By analogy with the structures shown above the corresponding growing cation in the polymerization of THF can be visualized as follows:

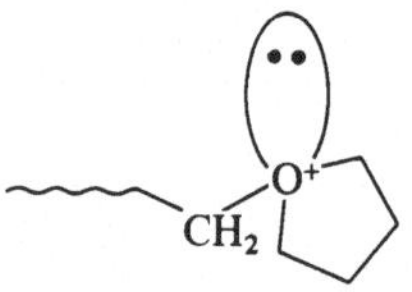

Studies of the distribution of the positive charge (electron deficiency) in oxonium ions (calculations using the CNDO-2 method) showed that the positive charge is located not on the oxygen atom but mainly on the adjacent carbon and hydrogen atoms. The calculated charge distribution in the trimethyloxonium ion may be illustrated as follows[66]:

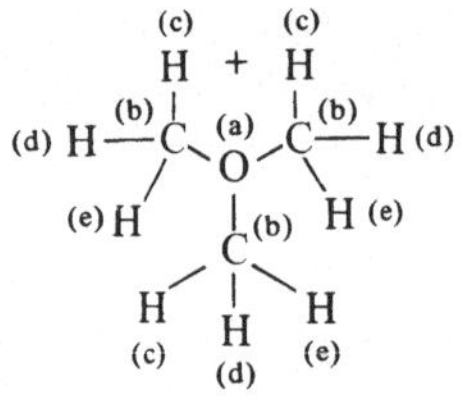

a	+0.038
b	+0.128
c, d, e	0.056, 0.068, 0.073
	(all three protons are non-equivalent)

The oxonium ions are isoelectronic with amines and both have similar structures. The activation energy for the inversion of the isopropyloxiranium cation ($\Delta E^{\ddagger} = 10 \pm 2$ kcal $\cdot$ mole^{-1})[163] is close to that of N-phenylaziridine ($\Delta E^{\ddagger} = 12$ kcal $\cdot$ mole^{-1})[166]. The angle between the aziridine ring and the phenyl substituent is $\phi = 125°$[167]. It agrees well with CNDO-2 calculations ($\phi = 124°$) for the methyloxiranium cation[165].

Sulfonium ions differ substantially in structure. The inversion barrier is much higher ($\Delta H^{\ddagger} = 20 \div 30$ kcal $\cdot$ mole^{-1}) and in some cases optically active isomers having a high stability were separated (e.g. Me, Et, n-Bu $-$ S^{+})[168]. The bond angles are much smaller and much closer to 90°. With increasing size of the electronic cloud around the heteroatom, the hybridization of the free electron pair with those involved in chemical bonding is of less importance and bond angles approach 90° e.g.:

H_2O	104° 30′		NH_3	107°		NMe_3	108°
H_2S	92°		PH_3	93° 30′		PMe_3	100°
H_2Se	90°		AsH_3	92°		$AsMe_3$	96°

The importance of the influence of substituents also decreases with increasing size of the atom.

4.2 Macroions and Macroion Pairs in Propagation

4.2.1 Determination of the Concentration of the Growing Species [P*]

Two types of methods were used in the cationic polymerization of heterocycles to determine the concentration of the growing species. The first type is based on methods used in free-radical polymerization (radical trapping) and in anionic polymerization when the macroanions are quantitatively converted into the stable end groups whose concentrations can then be measured.

In cationic polymerization end-capping was first applied by Saegusa in the polymerization of THF[16] and then of other cyclic ethers[170]; sodium phenoxide was used as the end-capping reagent and the phenoxy end groups were determined by UV. End-capping is sufficiently fast and the phenolate is exclusively formed in reaction with the growing species:

$$\ldots -CH_2-{}^+O \bigpentagon + Na\, OC_6H_5 \longrightarrow \ldots -CH_2-O-(CH_2)_4OC_6H_5 + NaA \qquad (78)$$
$$A^-$$

The extinction coefficient of the macromolecular ether is $\epsilon = 1.93 \cdot 10^3\, mole^{-1} \cdot l \cdot cm^{-1}$ at $\lambda_{max} = 272$ nm[169].

This method has recently been modified by using a picrate; the corresponding covalent picrates have much higher extinction coefficients ($\epsilon = 1.36 \cdot 10^4\, mole^{-1} \cdot l \cdot cm^{-1}$ at $\lambda_{max} = 250$ nm) than the phenolate[171]. However, picrate anions are known to be much weaker nucleophiles and it is not clear at present whether termination (end-capping) is sufficiently fast.

Although end-capping methods allow the concentration of the growing species to be measured, but no information about their chemical structures can be obtained in this way.

It has recently been shown[157] that the growing macrocations react rapidly with triaryl- or trialkylphosphines. This process, called "ion trapping" by analogy with radical trapping, gives stable quaternary phosphonium ions, e. g.:

$$\ldots -CH_2-{}^+O \bigpentagon + PR_3 \longrightarrow \ldots -CH_2-O-(CH_2)_4-\overset{+}{P}R_3 \qquad (79)$$

The concentration of phosphonium ions can be measured by FT-^{31}P-NMR; reliable results were obtained even for concentrations of the growing species as low as $10^{-4}\, mole \cdot l^{-1}$ [157]. The ^{1}H-NMR method has been applied to the determination of the structure of the growing species (cf. Table 7) and it can also be used (particularly when FT technique is available) for the determination of their concentrations.

4.2.2 Dissociation of Macroion Pairs

As in case of other ionic reactions, onium ions in the polymerization of heterocycles can exist in various forms of ionic aggregates, as paired or free ions, etc.

Detailed studies of the dissociation of macroion pairs, including thermodynamic parameters of dissociation, are available only for THF[172, 173] and oxepane[174]. Dissociation constants K_D and the corresponding enthalpies and entropies of dissociation are given in Table 8 below.

Measurements of K_D were performed for polymeric chains differing in $\overline{DP}_n$, but it has recently been shown for the macromolecular ammonium salts

$$\wwwww\overset{+}{N}R_4\, SbCl_6^-$$

that K_D is independent of chain length[175].

Table 8. Dissociation constants K_D and thermodynamic parameters (ΔH_D and ΔS_D) of dissociation of macroion pairs

Monomer	$[M]_0$ (mole $\cdot$ l^{-1})	Solvent	Dielectric constant of solution	K_D (mole $\cdot$ l^{-1})	ΔH (kcal $\cdot$ mole^{-1})	ΔS (cal $\cdot$ mole$^{-1} \cdot$ K^{-1})	Ref.
THF	7.0	CH_2Cl_2	8.2	$1.5 \cdot 10^{-5}$	-2.9 ± 1.3	-32 ± 5	[173]
	–	CH_3NO_2	22.8	$2.0 \cdot 10^{-3}$	-3.8 ± 0.6	-25 ± 2	[172]
Oxepane	5.0	CH_2Cl_2	7.4	$2.8 \cdot 10^{-5}$	-0.9 ± 0.4	-23.5 ± 1.5	[174]
		$C_6H_5NO_2$	19.1	$1.58 \cdot 10^{-3}$	-1.7 ± 0.1	-18.7 ± 0.5	

Figure 4 describes the dependence of ln K_D for poly-THF and polyoxepane on the reciprocal of the bulk dielectric constant.

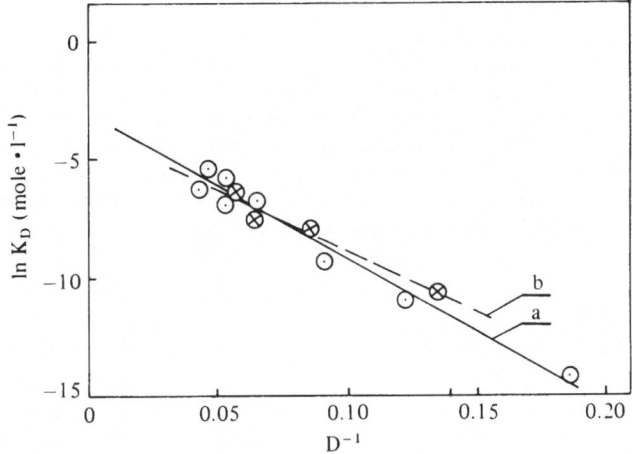

Fig. 4. Dependence of ln K_D for poly-THF (*a*) and poly-OXP (*b*) macroion pairs on the reciprocal of the bulk dielectric constant D at 25 °C

The dissociation of poly-THF macroion pairs in THF/solvent mixtures is more exothermic than the dissociation of macroion pairs of polyoxepane under similar conditions (dielectric constant, solvent structure, temperature). In both cases, however, the exothermicity is very low and typical of separated rather than of contact ion pairs. It has to be remembered that the positive charge ist not concentrated on the oxygen atom but distributed on the three carbon atoms and that, moreover, the anions with their diffused charge are large. The estimated dimensions of the poly-THF growing ion pair are shown below for SbF_6^- anion:

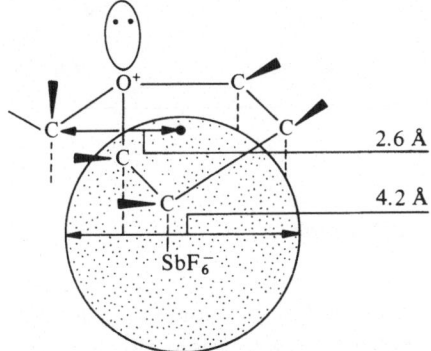

The small differences in ΔH_D and ΔS_D between poly-THF and polyoxepane macroion pairs in CH_2Cl_2 solvent (the dissociation of poly-THF macroion pairs is more exothermic and accompanied by a higher loss of entropy) stem from the higher nucleophilicity of THF. THF, as a

component of the medium around the ions, provides more ordered structure in the vicinity of the cation (oxepane exhibits a puckered structure while THF is practically a flat molecule).

The ΔH_D and ΔS_D values of triphenylmethylium salts are also close to those found for poly-THF and polyoxepane ion-pairs, being equal to -2.2 kcal $\cdot$ mole^{-1} and -25.0 e.u., respectively (measured for AsF_6^- and SbF_6^- anions in CH_2Cl_2 at 25 °C)[45].

A change of the solvent polarity (from CH_2Cl_2 to CH_3NO_2 or $C_6H_5NO_2$), causes a 10^2 fold increase in K_D of both the poly-THF and polyoxepane ion pair. It is difficult to judge whether this effect is mainly due either to a change in the polarity at the site of the ion-pair location (according to the Denison-Ramsey formula[177]) or to the increased solvation. Thus, the large increase of K_D for the poly-THF macroion pair when passing from e.g. CH_2Cl_2 to CH_3NO_2 indicates that in the immediate vicinity of the ion pair there is room for both monomer and solvent molecules, in spite of the higher nucleophilicity of the monomer molecules.

The values of K_D found in CH_2Cl_2 solution for different monomers or model systems do not differ too much from each other (Table 9).

Table 9. Dissociation constants for different model systems of active species in CH_2Cl_2

Salt	Temp. (°C)	$K_D \cdot 10^5$ (mole $\cdot$ l^{-1})	Ref.
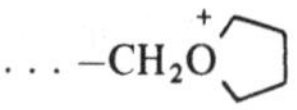 $\ldots -CH_2O^+$ ⟨ring⟩ , SbF_6^-	0	3.0	173)
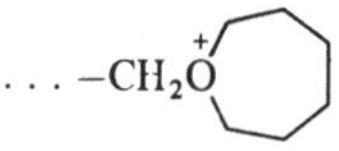 $\ldots -CH_2O^+$ ⟨ring⟩ , SbF_6^-	0	3.1	174)
$EtOCHCH_3^+, SbCl_6^-$	−6	1.0	176)
$Et_4N^+, SbCl_6^-$	0	8.4	175)
$Et_3C_{16}H_{33}N^+, SbCl_6^-$	0	7.2	175)
Et_3S^+, BF_4^-	+20	1.3	94)
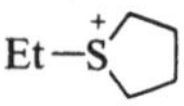 $Et-S^+$ ⟨ring⟩ , BF_4^-	+20	4.5	94)
Et_3O^+, PF_6^-	0	0.8	150)

4.2.3 Propagation on Macroion Pairs with Different Anions

The structure of the cation (counterion) in the anionic polymerization of e. g. styrene is of paramount importance for the kinetics of propagation. This was amply demonstrated by the works of Szwarc[8], Schulz[9], Bywater[10], a. o.

Cations used in anionic polymerization markedly differ in their properties. The small Li^+ has an ionic radius equal to 0.60 while Rb^+ or Cs^+ have radii equal to 1.69 A and 1.98 A, respectively[178].

The choice of counterions (anions) in the cationic polymerization of heterocyclic monomers can be almost as wide as in anionic polymerization, but only for the most nucleophilic monomers (i. e. cyclic amines). Unfortunately, in the polymerization of cyclic ethers, this choice is much more restricted. Thus, the small anions like F^- or OH^- cannot be used because, due to their high nucleophilicity and ability to form covalent bonds, they give rise to fast termination. In order to suppress or even to eliminate termination by collapse within an ion pair (cf. Sect. 5.1.), it is necessary to use complexed anions having large ionic radii. These are shown below (r_{cryst}):

Anion	BF_4^-	PF_6^-	SbF_6^-	$SbCl_6^-$	$CF_3SO_3^-$	ClO_4^-	AsF_6^-
Ionic radius (Å)	$2.30^{179)}$	$2.60^{179)}$	$2.10^{45)a}$	$3.00^{180)a}$	$2.96^{181)}$	$2.40^{181)}$	$2.10^{45)a}$

a   Stoke's radius

There are only a few systematic studies of the influence of anions on the propagation rate constant of macroion pairs[40, 125, 172]. In the polymerization of THF it was found that the propagation rate constants of various macroion pairs are independent of the structure of the anion. Below some pertinent data are given for different anions ($[THF]_0 = 7.0$ mole $\cdot$ l^{-1} at 25 °C in CH_3NO_2[172]).

Anion	AsF_6^-	$CF_3SO_3^-$	SbF_6^-	FSO_3^-
$k_p^{\pm}$ (mole$^{-1} \cdot$ l $\cdot$ s$^{-1} \cdot 10^2$)	2.02	2.05	2.1	2.21

In an earlier work, similar qualitative observations were reported[182].

The identity of k_p for various anions has three major origins: the large size of the anions, their similarity in structures and in solvation. Besides, this identity can also originate in some features of the macrocations, namely:
a) macrocations are large with diffused charges,
b) macrocations are probably specifically solvated by monomer, making macroion pairs even more similar, in spite of some difference in the anions.

In the polymerization of 3,3-dimethylthietane, $k_p^{\pm}$ for BF_4^- and $SbCl_6^-$ anions were also reported to be almost identical: $k_{(BF_4^-)} = 1.8 \cdot 10^{-3}$, $k_{(SbCl_6^-)} = 2.8 \cdot 10^{-3}$ (in mole$^{-1} \cdot$ l $\cdot$ s^{-1} at 20 °C in CH_2Cl_2)[183].

Enikolopyan studied the polymerization of cyclic amines such as conidine:

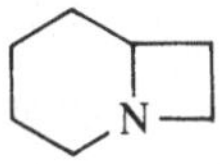

Conidine
(1−azobicyclo[4, 2, 0]octane)

with Cl^- (1.81), Br^- (1.96), J^- (2.15), and ClO_4^- (2.36) anions (ionic radii in Å given in brackets)[40]. Rate constants of propagation of ion pairs were found to increase with the size of the anion (the smallest for Cl^- and the largest for ClO_4^-). Values of $k_p^{\pm}$ were calculated from the kinetic treatment of the system propagating by free ions and ion pairs, whereas the degrees of dissociation were not directly taken from the studies of dissociation of the macroion pairs but from the K_D values of the respective initiators (N-ethylconidinium salts). Moreover, the range of reactivities given above may suggest that a fraction of the supposed ions existed in fact as unequally reactive covalent species. This has to be verified because, on the other hand, this system seems to be very suitable for the studies of the kinetics of propagation; its living nature was shown without any ambiguity.

4.2.4 Reactivities of Macrocations and Macroion Pairs

There are two groups of reports describing the determination of the rate constants of propagation of macrocations (k_p^+) and macroion pairs ($k_p^{\pm}$) in the polymerization of heterocycles. In the first one, k_p^+ and $k_p^{\pm}$ were found to be indistinguishable and in the second group, k_p^+ and $k_p^{\pm}$ differ a few times.

The determination of the small differences between k_p^+ and $k_p^{\pm}$ requires an extremely careful treatment of the experimental data. Numerical values of k_p^+ and k_p^+ are determined from the dependence of the observed (apparent) second-order rate constants (k_p^{app}) on the degree of dissociation of the macroion pairs. The most general treatment is based on the application of the simple relationship

$$k_p^{app} = \alpha \, k_p^+ + (1 - \alpha) \, k_p^{\pm} \tag{80}$$

being equivalent to:

$$k_p^{app} = k_p^{\pm} + \alpha \, (k_p^+ - k_p^{\pm}) \tag{81}$$

Thus, plotting k_p^{app} against α, one gets $k_p^{\pm}$ from the intercept and $(k_p^+ - k_p^{\pm})$ from the slope of the resultant straight line.

Another experimental approach involves application of Szwarc's plot derived from Eq. (80) and in which k_p^{app} is plotted against [living ends]$^{-1/2}$ ($k_p^{app} = k^{\pm} + k^- \cdot K_D^{1/2} \, [LE]^{-1/2}$)[184]. In the derivation of this equation it has been assumed that the rate constant involving macroions is much higher than that of the macroion pairs and that the concentration of macroions is much lower than that of the macroion pairs.

The application of Eq. (81) requires that k_p^{app} is determined for the precisely known concentration of the growing species. Sometimes, however, initiation is slow and at higher starting initiator concentration ([initiator]$_0$), the equality [initiator]$_0$ = [living ends] is not fulfilled when measurements are performed. On the other hand the degree of dissociation, which changes with starting conditions, has to be known for the prevailing conditions. Thus, one has to ensure both the reliable determination of the concentration of living ends and to know K_D (from which α is then calculated) of the macroion pairs for the prevailing conditions.

Change of α, necessary to realize a plot of Eq. (81), can be achieved either by changing the starting concentration of the initiator or by influencing K_D in some other way.

If the studied polymerization proceeds to equilibrium, then the equilibrium monomer concentration [M]$_e$ is included into the kinetic equation. However, [M]$_e$, for real solutions, changes with the initial concentration of monomer [M]$_0$ and may also vary (for short chains) with [living ends]$_0$. This has to be taken into account in order to eliminate a possible error.

Another difficulty arises when termination is accompanied by propagation (e.g. in the case of cyclic sulfides). It is not enough to kinetically extract k_p^{app} from the kinetic data in these systems but it has also to be shown that K_D does not change as polymerization proceeds. Actually, new kinds of ion pairs are formed in the termination step which may influence K_D and thus α.

All previous considerations mainly relate to measurements of k_p^+ and $k_p^\pm$; unfortunately, the number of works on the cationic polymerization of heterocyclic monomers fulfilling the requirements discussed above are very limited while several of the pitfalls described have not been avoided in some published papers.

Polymerization of oxepane initiated by 1,3-dioxolenium salts with SbF_6^- anions in CH_2Cl_2 and $C_6H_5NO_2$ solution provides a system devoid of the difficulties discussed above: the rate constant of initiation is 10^2 times higher than k_p, the invariable concentration of the growing species was determined by the "ion-trapping" ^{31}P-NMR method ([living ends] = [initiator]$_0$ within 5%), the propagation is sufficiently slow to allow measurements of K_D for various polymerization degrees thus ensuring the independence of K_D relative to $\overline{DP}_n$, and $[M]_e$ is low ($[M]_e = 0.08$ mole $\cdot$ l^{-1}).

The plot according to Eq. (81) (k_p^{app} vs. α)[174] is shown in Fig. 5. The upper line gives k_p^{app} in CH_2Cl_2 and the lower one k^{app} in $C_6H_5NO_2$. Analysis of these lines by the least-squares method gives the following results (at 25 °C):

CH_2Cl_2: $\quad k_p^\pm = (6.0 \pm 0.9) \cdot 10^{-4}$ mole^{-1} $\cdot$ l $\cdot$ s^{-1}

$\qquad\qquad k_p^+ - k_p^\pm = 0.2 \cdot 10^{-4}$ mole^{-1} $\cdot$ l $\cdot$ s^{-1}

$C_6H_5NO_2$: $\quad k_p^\pm = (4.7 \pm 0.7) \cdot 10^{-4}$ mole^{-1} $\cdot$ l $\cdot$ s^{-1}

$\qquad\qquad k_p^+ - k_p^\pm = -0.6 \cdot 10^{-4}$ mole^{-1} $\cdot$ l $\cdot$ s^{-1}

Thus, the differences between k_p^+ and $k_p^\pm$ are within the experimental error of kinetic measurements.

In the polymerization of THF studied in CH_2Cl_2[173] and CH_3NO_2[172] with AsF_6^- and SbF_6^- anions, it has also been found that $k_p^+ = k_p^\pm$ within the experimental error.

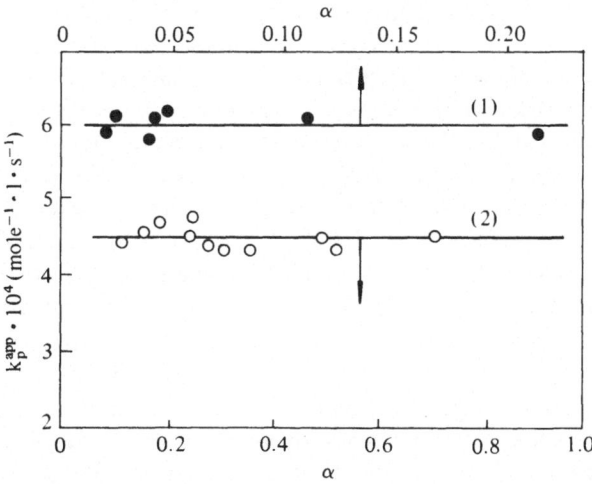

Fig. 5. Determination of k_p^+ and $k_p^\pm$ in the polymerization of oxepane (OXP). k_p^{app} as function of the degree of dissociation ofenacroion pairs: α; $[OXP]_0 = 5.0$ mole $\cdot$ l^{-1}, 25 °C. (1) CH_2Cl_2 as solvent, (2) $C_6H_5NO_2$ as solvent (Ref. 174)

There are, however, two other reports on the reactivities of free and paired ions in the polymerization of THF [185, 186] where macrocations are claimed to be more reactive than macroion pairs. The possible reasons for misinterpretation of the experimental data in these works are given in Ref. 173.

The polymerization of cyclic sulfides and amines provided, although tentatively, some information on k_p^+ and $k_p^\pm$. The equality $k_p^+ \approx k_p^\pm$ was found (the K_D value measured for an initiator used was further applied in the determination of α in polymerization studies) in the polymerization of 3,3-dimethylthietane [94] in $C_6H_5NO_2$ and qualitatively for 1-phenylmethyl-2-methyl-aziridine [187]. Small differences were found between k_p^+ and $k_p^\pm$ (the macrocation being less reactive than the macroion pair) in the polymerization of conidine [40]. In this system, α was also determined from K_D measured for initiators and not for the actual macromolecular growing species.

Recently, a series of works have been published on the cationic polymerization of lactones (e.g. β-propiolactone and ϵ-caprolactone [188,189]) and various ionic species have been reported together with elaborate kinetic treatments and some electrochemical measurements. In our opinion the chemical structure of the growing species in the cationic polymerization of lactones has not yet firmly been established (see also Ref. 190) and, therefore, a more detailed discussion of these interesting and important systems must wait until these structures are known.

The polymerization of 1,3-dioxolane provides only qualitative information about the reactivities of various ionic species. Polymerization in CH_2Cl_2 and CH_3NO_2 solvents proceeds with almost the same apparent second-order rate constants, indicating that growing species of various ionic structures may have similar reactivities (the dissociation constant is almost 10^3 times higher for poly-THF ion pairs in CH_3NO_2 than in CH_2Cl_2 and one could expect a similar dependence for poly-DXL ion pairs). An attempt to determine k_p^+ and $k_p^\pm$ was made including, however, a number of assumptions [191].

Summarizing the available information on the reactivities of macrocations and macroion pairs, it seems well established that in the polymerization of THF [172, 173], oxepane [174], 3,3-dimethylthietane (in $C_6H_5NO_2$) [183], and 1-phenylmethyl-2-methyl-aziridine [187] k_p^+ and $k_p^\pm$ have the same values (within experimental error, estimated e. g. in the polymerization of oxepane to be about $\pm10\%$). These polymerizations were studied with large anions like AsF_6^-, $SbCl_6^-$, SbF_6^-, or BF_4^-. In other reports, giving various and less reliable results for reasons discussed above, small differences between k_p^+ and $k_p^\pm$ have been noted. The most notable case is the polymerization of conidine, the only polymerization studied with anions considerably differing in size. Studies of cyclic amines should in the future answer the question whether the present observation of $k_p^+ = k_p^\pm$ is primarily due to the large size of the majority of the anions studied and to the resulting weak electrostatic influence on the cation (being a site of nucleophilic attack). Thus, it has to be ascertained whether this observation is due to this forced choice or if it stems from the more general feature of the cationic polymerization of heterocycles where the specific solvation of macrocations and macroion pairs by highly nucleophilic monomers (or polymer segments) decreases the inherent differences in reactivities of these different ionic species.

There is another question that has to be discussed: the direction of an attack in the S_N2 reaction vs. the position of the anion in the onium salts. Indeed, the simple and attractive picture proposed by Szwarc in order to explain the differences between the reactivities of macroanions and macroion pairs in the anionic homopropagation of styrene is based on the assumption that an ion pair has to dissociate partially when the transition state of propagation is reached [192]. Szwarc, after observing a similar reactivity of the polystyryl anion and polystyryl cesium ion pair, also assumed that no partial dissociation was needed for the large Cs^+ cation. The first part of this reasoning tacitly assumes that the stereochemistry of the ion pair is such that partial dissociation is really needed, i.e. that the two electrons contributing to the pairing with the cation should be available for the electrophilic attack by an approaching monomer molecule.

Let us examine the $S_N 2$ reaction involving an onium ion, e.g. constituted by an ethylene oxide moiety:

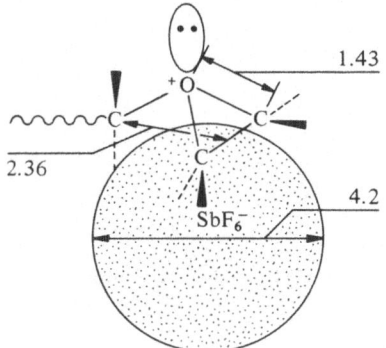

(the shaded area denotes the anion; distances are given in Å)

Nucleophilic attack of the incoming monomer molecule is directed along the $C - \overset{+}{O}$ bond in the oxonium ion and does not require a marked rearrangement of the ion pair; the anion has merely to be slightly displaced toward the new oxonium ion being formed when a bond between a carbon atom being attacked and an oxygen atom in the form of an oxonium ion breaks. This is an important difference from anionic vinyl systems, in which the counterion, as discussed above, has to be removed much further away from the growing ion.

Thus, the three factors: solvation, large size of anions and the resulting weak interaction with the cation, as well as the stereochemistry of attack on the onium ion seem to be responsible, either together or separately, for two facts discussed in the Sect. 4.2.4: the identity of $k_p^{\pm}$ for various anions and the close relationship if not identity of k_p^{+} and $k_p^{\pm}$ in the cationic polymerization of heterocyclic monomers.

4.2.5 Solvent Effects (Polarity, Solvation of Ions, Monomer Complexes) in Propagation

4.2.5.1 Equilibria and Properties of Media

The best representation of the influence of solvent on the cationic polymerization of heterocyclic monomers comes from the observation of changes in the equilibrium monomer concentration for polymerizations conducted in various solvents. We shall examine here two systems: the polymerization of THF and 1,3-dioxolane.

The dependence of the equilibrium monomer concentration $[THF]_e$ on the initial monomer concentration $[THF]_0$ in various solvents is shown in Fig. 6.

Without going into the details of the thermodynamic description of non-ideal systems (cf. Ivin's comprehensive review[193]), it is apparent from Fig. 6 that in the

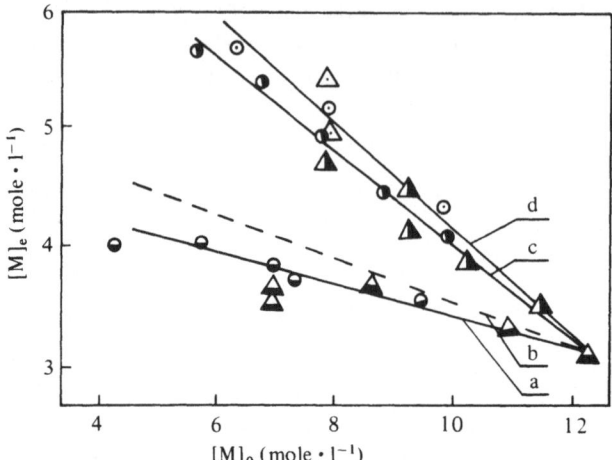

Fig. 6. Dependence of the equilibrium monomer concentration $[THF]_e$ on the initial monomer concentration $[THF]_0$ in $CCl_4(a)$, $C_6H_6(b)$[227], $CH_2Cl_2(c)$, $CH_3NO_2(d)$ at 25 °C (Ref. 13)

more acidic solvents higher equilibrium monomer concentrations are observed. Acidic (toward THF) solvents form with THF loose complexes (e.g. the enthalpy of mixing, ΔH, of THF with CCl_4 is equal to -0.7 kcal $\cdot$ mole^{-1} [194]) and with CH_2Cl_2 as much as -1.2 kcal $\cdot$ mole^{-1} [194]. Thus, on the molecular level, the observed dependences imply that the actual momentary concentration of THF available for polymerization (its activity in the thermodynamic sense) is lowered by the fraction of complexed monomer, provided that the complexed monomer propagates less rapidly.

1,3-Dioxolane is much less nucleophilic than THF and, therefore, the dependence of the equilibrium monomer concentration $[DXL]_e$ on the initial monomer concentration $[DXL]_0$ and solvent structure, although clearly demonstrated by Enikolopyan[195], is much less pronounced than in the polymerization of THF.

A knowledge of the dependence of $[M]_e$ on $[M]_0$ is of primary importance in the studies of the kinetics of polymerization. The use of the wrong value of $[M]_e$ may lead to erroneous values of k_p, especially when polymerization is studied where $[M]_0$ is close to $[M]_e$.

4.2.5.2 Activation Parameters of Propagation and Properties of Media

The dependence of $\Delta H_p^{\ddagger}$ and $\Delta S_p^{\ddagger}$ on the properties of the medium, namely solvent structure, its solvation power and polarity, was studied for the ionic propagation of THF.

Figure 7 depicts the dependence of $\Delta H_p^{\ddagger}$ and $\Delta S_p^{\ddagger}$ on the properties of the medium as a compensation plot ($\Delta H_p^{\ddagger}$ plotted against $\Delta S_p^{\ddagger}$) and Fig. 8 the dependence of k_p^i ($k_p^i = k_p^+ = k_p^{\pm}$; cf. Sect. 4.2) on D. It is remarkable that k_p^i changes only slightly (e. g. from $k_p^i = 2.1 \cdot 10^{-2}$ mole$^{-1} \cdot$ s^{-1} in CH_3NO_2[173]) to $k_p^i = 4.4 \cdot 10^{-2}$ mole$^{-1} \cdot$ s^{-1} in CCl_4[125]; both at $[THF]_0 = 7.0$ mole $\cdot$ l^{-1} at 25 °C) whereas $\Delta H_p^{\ddagger}$ and $\Delta S_p^{\ddagger}$ separately change immensely. It is convenient to analyze two sets of results. In the first one, the initial concentration of

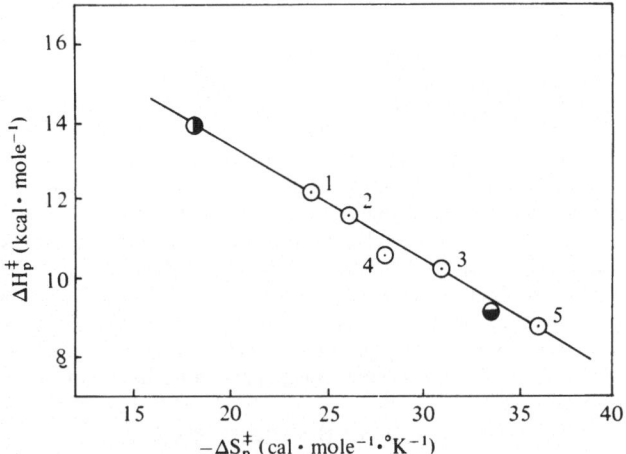

Fig. 7. Isokinetic plot for propagation of THF; ⊙ $[THF]_0$ = 7.0 mole $\cdot$ 1^{-1} (polarity of the CH_3NO_2/CH_2Cl_2 mixtures decreases 1 → 5) ◕ $[THF]_0$ = 8.0 mole $\cdot$ 1^{-1}, CCl_4 solvent ◐ $[THF]_0$ = 12.3 mole $\cdot$ 1^{-1} (Ref. 173)

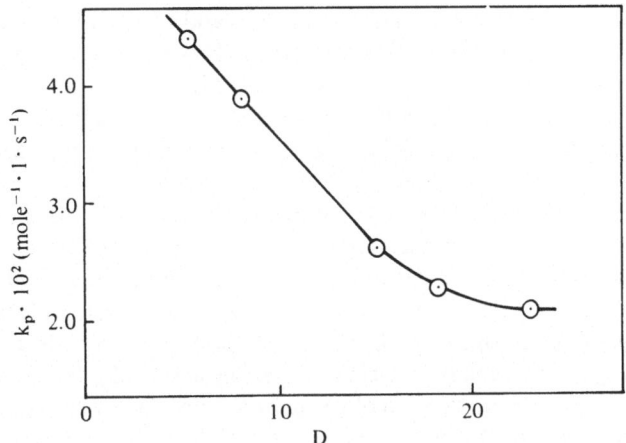

Fig. 8. Dependence of k_p^i on the dielectric constant at 25 °C, $[THF]$ = 7.0 mole $\cdot$ 1^{-1}

monomer is kept constant ($[THF]_0$ = const.) and the solvent is changed; CCl_4, CH_2Cl_2, CH_3NO_2 and their mixtures were used as solvents. In this set of experiments $\Delta H_p^{\ddagger}$ increases and $\Delta S_p^{\ddagger}$ decreases (compensating the change of enthalpy). The value of k_p^i, however, monotonously decreases with solvent polarity. If, however, a mixture of THF/CH_2Cl_2 is studied where polarities (expressed by dielectric constants) are very close to each other (THF: ϵ = 7.6 and CH_2Cl_2: ϵ = 9.0; both at 25 °C), then, with increasing $[THF]_0$ (better solvating component), $\Delta H_p^{\ddagger}$ increases and $\Delta S_p^{\ddagger}$ decreases but (at 25 °C) k_p^i remains almost constant.

These results can be summarized as follows:

A. $[THF]_0$ = const., polarity (ϵ) = variable (CCl_4, CH_2Cl_2, and CH_3NO_2 were used)

polarity (ϵ) ↑: $\Delta H_p^{\ddagger}$ ↑; $\Delta S_p^{\ddagger}$ ↓ ; k_p^i ↓

B. $[THF]_0$ = variable; polarity (ϵ) = const. (THF/CH$_2$Cl$_2$ system)

$[THF]_0 \uparrow$; $\Delta H_p^{\ddagger} \uparrow$; $\Delta S_p^{\ddagger} \downarrow$; $k_p^i \sim$ const.

Thus, $\Delta H_p^{\ddagger}$ and $\Delta S_p^{\ddagger}$ are governed by the solvation power of the medium whereas $\Delta G_p^{\ddagger}$ (= $\Delta H_p^{\ddagger} - T\Delta S_p^{\ddagger}$), at temperatures close to 25 °C is independent of solvation power. This is an artifact; it just so happened that at temperatures close to 25 °C, $\Delta\Delta H_p^{\ddagger} = \Delta\Delta S_p^{\ddagger} \cdot T$.

The polarity expressed by the dielectric constant has some, although small, influence on the activation parameters, as indicated by set A above.

THF is the most powerful solvating agent in the systems studied (CCl$_4$, CH$_2$Cl$_2$ and CH$_3$NO$_2$); however, changing (for $[THF]_0$ = 7.0 mole $\cdot$ l^{-1}) from CCl$_4$ (ϵ = 5.9) to CH$_3$NO$_2$ (ϵ = 20.6) leads to a small decrease of the measured k_p^i value. These facts mainly imply that the closest surroundings of the growing macrocations (or macroion pairs) are not merely composed of clusters of THF molecules and polymer fragments as suggested by Szwarc[196] but that the solvent, even the non-polar one, deeply enough penetrates the site of the reaction to influence all kinetic parameters.

The present theories of the effects of solvents on the rates of polar (ionic) reactions do not permit a quantitative analysis of the above cited results. Fractions of $\Delta H_p^{\ddagger}$ and $\Delta S_p^{\ddagger}$, due to solvation, apparently compensate each other, because the increase in the energy needed for desolvation is just compensated by equal contributions of the entropy (a more firmly bound or larger number of molecules are desolvated). This compensation phenomenon is well known in organic chemistry. For instance, the difference between the activation enthalpies ($\Delta\Delta H^{\ddagger}$) of the reaction of benzyl chloride with pyridine in DMF and CH$_3$OH is equal to -5.3 kcal $\cdot$ mole^{-1}. The rate constants of this reaction are however nearly the same in both solvents (k = 3.46 $\cdot$ 10^{-6} mole$^{-1} \cdot$ l $\cdot$ s^{-1} (CH$_3$OH) and k = 3.74 $\cdot$ 10^{-6} mole$^{-1} \cdot$ l $\cdot$ s^{-1} (DMF), both at 25 °C). This is due to the fact that the change of activation enthalpy is compensated by the parallel change of activation entropy ($\Delta\Delta S^{\ddagger}$ = = 17.6 cal $\cdot$ K$^{-1} \cdot$ mole^{-1})[197].

Thus, the net influence of increased polarity on the rate constants is, in this case as well as in polymerization of THF, eventually not attributable to the solvation phenomenon itself but to the stronger electrical microfields formed by more easily induced dipoles in more polar (and more polarizable) solvent molecules (e.g. CH$_3$NO$_2$ vs. CCl$_4$). In this way, the energy of the ground state decreases and $\Delta G_p^{\ddagger}$ becomes larger in more polar solvents, decreasing eventually the corresponding rate constants.

The polymerization of oxepane has provided very similar results[174], Goethals has observed that in the polymerization of 3,3-dimethylthietane k_p^i is also larger in CH$_2$Cl$_2$ than in C$_6$H$_5$NO$_2$[94].

4.3 Rate Constants of Propagation and Structure of Monomers

In polymerization, as in any other complex chemical processes consisting of a number of elementary reactions, first of all methods have to be found to study these reactions separately.

Monomer is predominantly consumed in the propagation step, and if the concentration of the growing species can be determined, then even for the non-living systems there is a good chance to measure the reliable values of k_p by using simple relationship:

$$-d[M]/dt = k_p \cdot [P^*] \cdot [M]$$

where $[P^*]$ is the concentration of the growing species.

As discussed earlier (Sect. 4.2), the multiplicity of the forms of ionic species makes measurements of k_p of elementary reactions more complicated; nevertheless, these few cases in which k_p for macrocations and macroion pairs have been determined separately are also discussed in this section. It has been shown that in the majority of the systems studied $k_p^+ \approx k_p^\pm$ and, therefore, in Tables 10–12 listing the values of k_p, we have tentatively assumed that k_p represents the rate constants of propagation of macroions and macroion pairs which coincide unless stated otherwise.

In Tables 10–12, the values of k_p are listed for a number of heterocyclic monomers; these values were determined either on the bases of analytically determined concentrations of active species or with an assumption that the concentration of the growing species is equal to the initial concentration of initiator. This assumption led to a number of incorrect values of k_p listed in the literature, particularly for the polymerization of cyclic acetals. A comprehensive critical treatment was published on this subject[208], some of the reasons for the low efficiency of initiation are discussed in Sect. 3.

It has conclusively been shown by ^{1}H-NMR that in the polymerization of THF, chain transfer to polymer can be neglected, at least for lower conversion of monomer. Thus, all of the oxonium ions determined by end capping or by ion trapping are the growing tertiary oxonium ions. This reaction cannot, however, be excluded in other systems:

Cyclic, reactive "Dormant" species, less reactive (also in the poly- (81) merization of cyclic sulfides, amines and phosphates)

More recently, the application of ^{1}H-NMR confirmed that in the polymerization of oxepane the transfer to polymer (reaction (81)) has no kinetic importance[209]. There is, however, no evidence that in the polymerization of other cyclic ethers, particularly more strained ones like oxetane and 3,3-dimethyloxetane, a portion of tertiary oxonium ions is not in the form of the "dormant", non-reactive species. A few papers have reported the formation of larger rings, apparently by unimolecular back-biting, which may reduce the concentration of active species and eventually give wrong values of "k_p", if the method used for the determination of $[P^*]$ does not discriminate between two kinds of oxonium ions given in Eq. (81).

In order to make the comparisons possible for data provided by various authors we recalculated the values of k_p, whenever it was possible, by using the activation parameters and 0 °C as the standard condition.

Besides the data listed in Tables 10–12, there are also values of k_p we have not used. These values are either "too small" or "too large" in comparison with those of the majority of other authors. The possible reason for "too low" values in the polymerization of THF in one series of

Table 10. Rate constants of propagation and activation parameters in the polymerization of THF at 0 °C

No	$[M]_0$ (mole·l^{-1})	Initiator, $[I]_0 \cdot 10^2$ (mole·l^{-1})	Solvent	$k_p \cdot 10^3$ (mole^{-1}·l· s^{-1})	$\Delta H^{\ddagger}$ (kcal· ·mole^{-1})	$\Delta S^{\ddagger}$ (e.u.)	Ref.
1	12.6	Et$_3$O$^+$, BF$_4^-$, >1.0	bulk THF	3.3	12.7	−25.2	135)
2	12.6	 ⟨O$^+$⟩O, SbF$_6^-$, var	bulk THF	4.4	14	−18	173)
3	12.6	BF$_3$ · ECH, 1.0	bulk THF	4.5	–	–	198)
4	7.0	 ⟨O$^+$⟩O, SbF$_6^-$, var	CH$_3$NO$_2$	3.9	12	−24	172)
5	7.0	 ⟨O$^+$⟩O, SbF$_6^-$, var	CH$_2$Cl$_2$	5.7	10	−32	173)
6	6.1	Et$_3$O$^+$, BF$_4^-$, ~1.0	1,2-Cl$_2$C$_2$H$_4$	5.0	9.4	−34	128)
7	7.5	Et$_3$O$^+$, BF$_4^-$	CH$_2$Cl$_2$	4.8	–	–	78)
8	7.5	Et$_3$O$^+$, PF$_6^-$	CH$_2$Cl$_2$	4.2	–	–	78)
9	7.5	Et$_3$O$^+$, SbF$_6^-$	CH$_2$Cl$_2$	5.9	–	–	78)
10	6.3	BF$_3$ · THF, ECH, >1.0	CH$_2$Cl$_2$	4.1	11.4	−28	198)
11	8.0	 ⟨O$^+$⟩O, SbF$_6^-$, var	CCl$_4$	9.5	9	−36	125)

ECH = α-epichlorohydrin

Table 11. Rate coefficients[a] and rate constants of propagation (k_p) and activation parameters in the polymerization of cyclic ethers (except THF) at 0 °C

No	Monomer	$[M]_0$ (mole·l^{-1})	Initiator	$[I]_0 \cdot 10^2$ (mole·l^{-1})	Solvent	k_p (mole^{-1}·l·s^{-1})	$\Delta H^\ddagger$ (kcal·mole^{-1})	$\Delta S_p^\ddagger$ (e.u.)	Ref.
1	OXP,	2.9	BF$_3$·ECH,	5.7	CH$_2$Cl$_2$	$1.5 \cdot 10^{-5}$	17.4	−18	199)
2	OXP,	5.0	 SbF$_6^-$,	var^b	CH$_2$Cl$_2$	$1.0 \cdot 10^{-4}$	13	−29	174)
3	OXP,	5.0	 SbF$_6^-$,	var^b	PhNO$_2$	$0.3 \cdot 10^{-4}$	14.8	−24.5	174)
4	OXT,	3.1	BF$_3$·THF,	0.8	CH$_2$Cl$_2$	$0.14^{a,c}$	13.7	−12.1	200)
5	OXT,	0.5	BF$_3$·THF,	0.6	CH$_3$C$_6$H$_{11}$	$1.4^{a,c}$	10.7	−19	200)
6	3MOXT,	1.2	BF$_3$·THF,	0.8	CH$_3$C$_6$H$_{11}$	$5.3^{a,c}$	11.3	−15	200)
7	3,3dMOXT,	0.4	BF$_3$·THF,	3	CH$_3$C$_6$H$_{11}$	$18.6^{a,c}$	12.1	−8.5	200)
8	endo 2M7OH,	1.4	BF$_3$·ECH,	0.7	CH$_2$Cl$_2$	$4.0 \cdot 10^{-2}$	13.4	−15.7	201)
9	exo 2M7OH,	1.4	BF$_3$·ECH,	1.3	CH$_2$Cl$_2$	$4.3 \cdot 10^{-2}$	14.4	−11.6	202)
10	BCMO,	0.3	Al(i-Bu)$_3$·H$_2$O,	>1	C$_6$H$_5$Cl	0.8^c	5.4	−39	122)

a   rate coefficients (see text above)
b k_p measured at various $[I]_0$
c   extrapolated values

OXP — Oxepane	3,3dMOXT — 3,3-Dimethyloxetane
OXT — Oxetane	2M7OH — 2-Methyl-7-oxabicyclo[2.2.1]heptane
3MOXT — 3-Methyloxetane	BCMO — 3,3-Bis-(chloromethyl)oxetane

Table 12. Rate constants of propagation and activation parameters in the cationic polymerization of cyclic sulfides, amines and iminoethers at 0 °C

No	Monomer	[M] (mole·l⁻¹)	Initiator	$[I]_0 \cdot 10^2$ (mole·l⁻¹)	Solvent	k_p (mole⁻¹·l·s⁻¹)	$\Delta H^{\ddagger}$ (kcal·mole⁻¹)	$\Delta S^{\ddagger}$ (e.u.)	Ref.
1	Conidine	6.0	free ions		CH_3OH	$5 \cdot 10^{-4}$ a	10.8	−34	40)
2	ABH,	6.0	free ions		CH_3OH	$4.4 \cdot 10^{-3}$ a	8.5	−38	40)
3	QNL,	6.0	free ions		CH_3OH	$4.4 \cdot 10^{-10}$ a	18.6	−33	40)
4	1-Methylazetidine,	1.0	$Et_3\overset{+}{O}$, BF_4^{-},	3.0	CH_2Cl_2	$9 \cdot 10^{-4}$ a	13.6	−23.4	203)
5	TMA		$Et_3\overset{+}{O}$, BF_4^{-},	4.0	$PhNO_2$	$5.6 \cdot 10^{-8}$ a	18	−23.3	204)
6	BMA		$Et_3\overset{+}{O}$, BF_4^{-},	1.0	CH_2Cl_2	$1.2 \cdot 10^{-2}$			187)
7	Thietane				CH_2Cl_2	0.9b			205)
8	DMT				CH_2Cl_2	$1.1 \cdot 10^{-3}$	12.5	−26	205)
9	DET				CH_2Cl_2	$2 \cdot 10^{-4}$ b			205)
10	PhOxz		APOP,	3.0	$CH_3C(O)N(CH_3)_2$	$1 \cdot 10^{-9}$ a	26.8	−2.0	206)
11	PhOxz		APOP,	3	CH_3CN	$8 \cdot 10^{-9}$ a	22.6	−11	206)
12	Oxz		MTs,	10^2	CD_3CN	$7 \cdot 10^{-6}$ a	24.4	+7.5	207)
13	MOxz		MTs,	10^2	CD_3CN	$1 \cdot 10^{-6}$ a	18.5	−18.2	207)

a extrapolated to 0 °C
b +20 °C

ABH – 1-Azobicyclo[3.1.0]hexane
QNL – Quinuclidine (1-azobicyclo[2.2.2]octane)
TMA – 1,3,3-Trimethylazetidine
BMA – 1-Benzyl-2-methylaziridine
DMT – 3,3-Dimethylthietane
DET – 3,3-Diethylthietane

PhOxz – 2-Phenyl-2-oxazoline
Oxz – 2-Oxazoline
MOxz – 2-Methyl-2-oxazoline
APOP – 3-Alkyl-2-phenyl-2-oxazolinium perchlorate
MTs – Methyl p-toluenosulfonate

works[185, 186] (2.10^{-3} mole $\cdot$ 1^{-1} $\cdot$ s^{-1} in CH_2Cl_2 for $(C_2H_5O)_3O^+BF_4^-$ and PF_6^- at 0 °C) was previously discussed[173]. It seems that only part of the initiator reacts when k_p is measured because in this particular system $k_i \ll k_p$, and the equality $[I]_0 = [P^*]$ used by the authors to calculate k_p apparently is invalid.

The agreement between k_p values measured by various authors under similar conditions (solvent, temperature) is usually good, sometimes very good (e.g. $5.7 \cdot 10^{-3}$ mole^{-1} $\cdot$ 1 $\cdot$ s^{-1} in CH_2Cl_2 and $[THF]_0 = 7.0$ mole $\cdot$ 1^{-1} at 0 °C and $5.9 \cdot 10^{-3}$ mole^{-1} $\cdot$ 1 $\cdot$ s^{-1} in the same solvent and at 0 °C for $[THF]_0 = 7.5$ mole $\cdot$ 1^{-1}, cf. also No 6 in Table 10). The agreement between the activation parameters has also to be considered as good (cf. No 1–2, and No 5, 6, 11). Thus, the values listed in the table, and particularly No 2, 4, 5 and 11 can be accepted as the actual values of k_p under given conditions. One has to remember, however, that k_p in the polymerization of THF changes with the composition of the polymerization mixture as described in Sect. 4.2.5.

Since it is known that the importance of chain transfer to polymer rapidly decreases with increasing degree of substitution of the monomer[205] and since, on the other hand, there is no reason to expect that substituted oxetanes are much more reactive (in the form of the corresponding oxonium ions and/or monomers) than oxetane itself, the large difference in k_p between oxetane and 3,3-dimethyloxetane could arise from larger proportions of active species being converted into the "dormant" species of the former monomer (cf. thietanes and azetidines in Table 12). The instantaneous concentration of growing species decreased in this way could give lower values of k_p. Thus, we will not discuss the reported values of k_p and the corresponding supposed "activation parameters". Until it will become clear that the presence of the dormant species does not influence the measured values of the rate constants, we will continue to call the numbers obtained in this way rate coefficients (R.C.) or even refrain from quoting the pertinent values.

A number of measurements were made also for cyclic acetals, mostly 1,3-dioxolane. Although the measured k_p value can be lower than the actual value (e.g. if a too high concentration of the active species is taken for calculations), it is much more difficult to find reasons why the measured value could have been higher than the real value of k_p. Thus, the values measured for 1,3-dioxolane are approximately equal to $k_p \sim 10^2$ mole^{-1} $\cdot$ 1 $\cdot$ s^{-1} [159, 208] and for 1,3-dioxepane $k_p \sim 10^4$ mole^{-1} $\cdot$ 1 $\cdot$ s^{-1}. These are at least the lower limits of the actual k_p value and it is immediately apparent that these values are much higher than any k_p value measured for cyclic ethers and other heterocyclics.

The analysis of the propagation rate constants leads to the following generalizations:

1) The k_p value increases for a given group of heterocycles with increasing ring strain and with decreasing degree of substitution. Ring strain facilitates breaking of the bond between a carbon atom and heteroatom while increasing substitution hampers the approach of a new monomer molecule.

2) For approximately the same ring strain, k_p for homopolymerization changes in the order:

acetals > ethers > sulfides > amines

This order is opposite to the order of nucleophilicities, which one could expect to be in line with the order of reactivities in homopropagation. Apparently, bond scission in the macrocation involved is more advanced in the transition state than bond formation of the incoming monomer. This is a known feature of the "borderline S_N2" mechanism, which is on the edge of becoming an S_N1 substitution.

Thus:

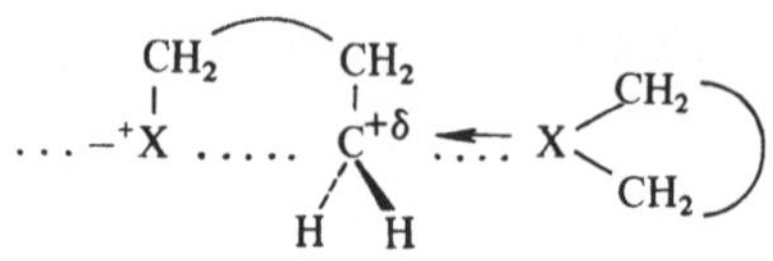

(Bond breaking more advanced ; X = heteroatom)

The higher the nucleophilicity of X, the more difficult becomes transfer of the electron defficiency (positive charge) to the leaving carbon atom (marked with $+\delta$). Obviously, in cyclic acetals, this transfer is facilitated because additional stabilization of the transferred positive charge is possible by the second adjacent oxygen atom. This oxygen atom is in α position to leaving positively charged carbon atom, being a site of the nucleophilic attack:

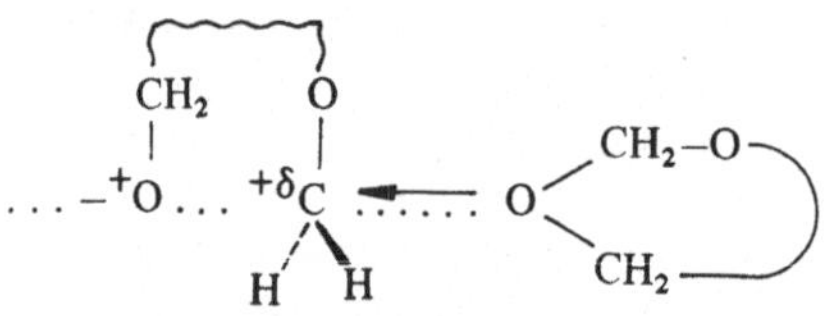

In order to compare the absolute reactivities of active species and monomers it is, however, necessary to react a series of monomers with a model active center and to react a series of active species with one chosen monomer. This, in principle, can be conducted in copolymerization studies, which are not included in this review. Nevertheless, we may cite here at least one work conclusively demonstrating that the rate constants of the addition of various monomers to a given active species do not change too much with monomer reactivity, most of the variations stemming from the diversity of the active species[205].

4.4 Stereochemistry of the Cationic Ring-Opening Propagation

4.4.1 Inversion of Configuration and Stereoselectivity

In the large majority of the cationic polymerizations of heterocycles the propagation step proceeds with complete inversion of configuration. This was shown for the first time by Vandenberg in the polymerization of cis- and trans-2,3-butene oxides[210] as well as for ethylene oxides substituted with deuterium[146].

Price cationically polymerized cis- and trans-dideuteroethylene oxides and analyzed the prepared polymers by IR spectra[211]. This work has recently been repeated and the microstructure of the chains was studied by the ^{13}C-NMR method (^{13}C satellites)[146].

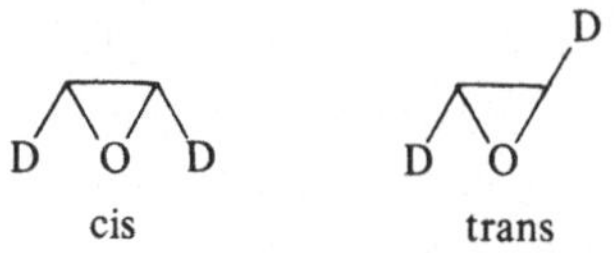

cis   trans

The polymerization of cis- and trans-monomers leads to the following chain configurations, depending on retention or inversion of configuration:

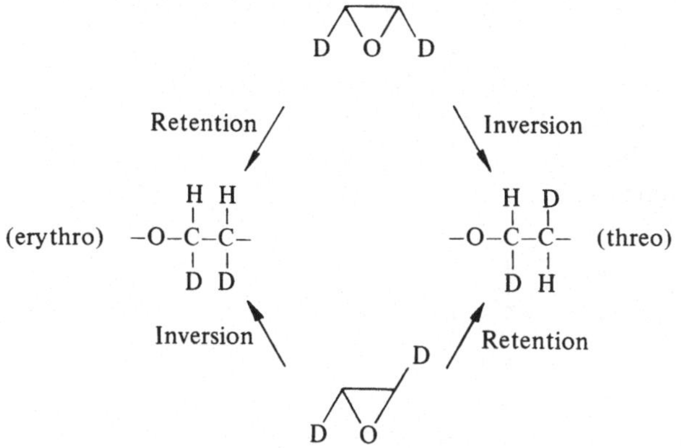

$$
\begin{array}{ccccc}
& & \overset{\triangle}{\underset{D\ \ O\ \ D}{}} & & \\
& \text{Retention}\diagup & & \diagdown\text{Inversion} & \\
& & & & \\
& \overset{H\ \ H}{\underset{D\ \ D}{-O-C-C-}} & & \overset{H\ \ D}{\underset{D\ \ H}{-O-C-C-}} & \\
\text{(erythro)} & & & & \text{(threo)} \\
& \text{Inversion}\diagdown & \underset{D}{} \diagup\text{Retention} & & \\
& & \overset{\triangle}{\underset{D\ \ O}{}} & &
\end{array} \tag{82}
$$

(erythro) $\underset{\overset{|}{D}\ \overset{|}{D}}{\overset{\overset{|}{H}\ \overset{|}{H}}{-O-C-C-}}$ $\underset{\overset{|}{D}\ \overset{|}{H}}{\overset{\overset{|}{H}\ \overset{|}{D}}{-O-C-C-}}$ (threo)

When a 70/30 mixture of cis- and trans-monomers is used, a 70/30 mixture of threo- to erythro-diads is obtained.

Thus, in agreement with the results observed by Price, who obtained polymers with a microstructure depending exclusively on the configuration of the monomers[211], ^{13}C-NMR studies of polymers reveal that polymerization of ethylene oxide almost entirely proceeds through inversion of configuration at the carbon atom attacked by a monomer molecule[146].

In the polymerization of cis- and trans-2,3-butene oxides, stereoregular polymers are obtained by a low-temperature cationic process[147]:

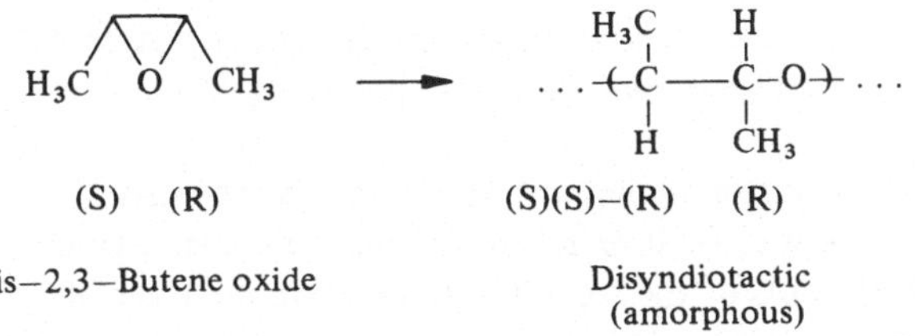

$$
\underset{\underset{(S)\quad\ (R)}{H_3C\ \ O\ \ CH_3}}{\overset{\triangle}{}} \longrightarrow \ \ldots \underset{\underset{(S)(S)-(R)\ \ (R)}{\overset{|}{H}\quad\overset{|}{CH_3}}}{\overset{\overset{H_3C}{|}\quad\overset{H}{|}}{+C-C-O+}}\ldots \tag{83}
$$

cis−2,3−Butene oxide   Disyndiotactic (amorphous)

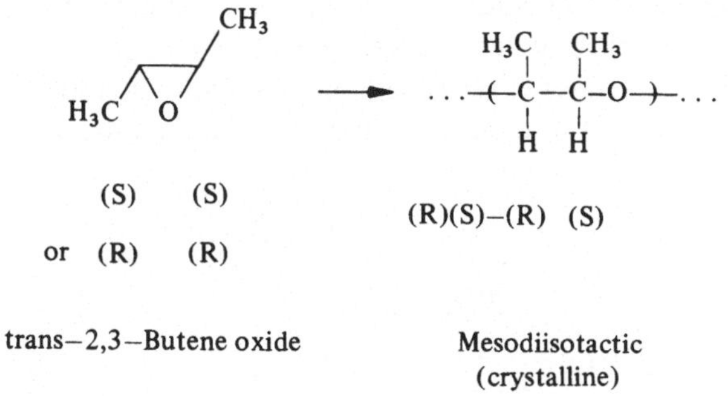

$$
\underset{\substack{(S)\quad\ (S)\\ \text{or}\ \ (R)\quad (R)}}{\overset{\overset{CH_3}{\diagup}}{\underset{H_3C\ \ O}{\triangle}}} \longrightarrow \ \ldots \underset{\underset{(R)(S)-(R)\ \ (S)}{\overset{|}{H}\ \overset{|}{H}}}{\overset{\overset{H_3C}{|}\ \overset{CH_3}{|}}{+C-C-O+}}\ldots \tag{84}
$$

trans−2,3−Butene oxide   Mesodiisotactic (crystalline)

The microstructure of these polymers was established by cleaving the corresponding polymers and determining the stereochemistry of the resulting glycols. The (RR, SS) disyndiotactic poly(cis-2,3-butene oxide) gives racemic 2,3-butanediol whereas (RS, RS) mesodiisotactic poly(trans-2,3-butene oxide) yields, on cleavage, meso 2,3-butanediol. The polymerization of trans-1,4-dichloro-2,3-butene oxide gives a polymer with a stereochemistry similar to that of poly(trans-butene oxide). However, the cationic polymerization of cis-1,4-dichloro-2,3-butene oxide as well as the cationic polymerization of 2,3-cis-butene sulfide lead to polymers differing in stereochemistry from poly(cis-2,3-butene oxide):

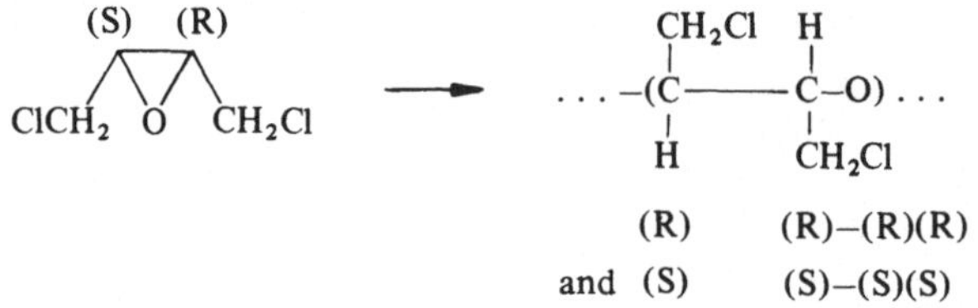

$$(85)$$

cis−1,4− Dichloro−
2,3− butene oxide

Racemic diisotactic
polymer

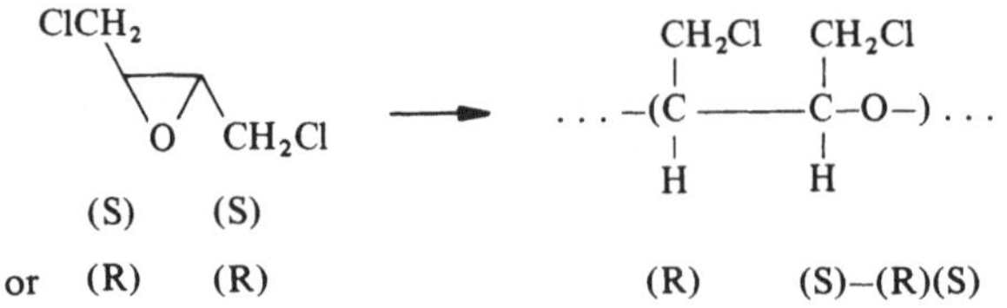

$$(86)$$

Mesodiisotactic

The stereochemistry of the polymers was established from their X-ray patterns and by a comparison of the derived (reduction with $LiAlH_4$) polymers with corresponding poly-2,3-epoxybutanes.

Vandenberg explains this variety of stereochemical structures on the basis of two kinds of interactions, namely the specific interactions between the last segment in the polymer chain and the substituents of the oxonium ion and on the basis of the steric interaction of the incoming monomer molecule and the oxonium ion.

Because the stereochemistry of polymers is independent of the initiator used, the stereoregulating effect of the counterions was excluded.

For trans-2,3-butene oxide and trans-1,4-dichloro-2,3-butene oxide, in order to understand the formation of disyndiotactic polymers, it is sufficient to assume a stereoselective process, i. e. that the active species selects a monomer with a stereochemistry resembling its own: Two different notations are used in order to conform with original papers

e.g.:

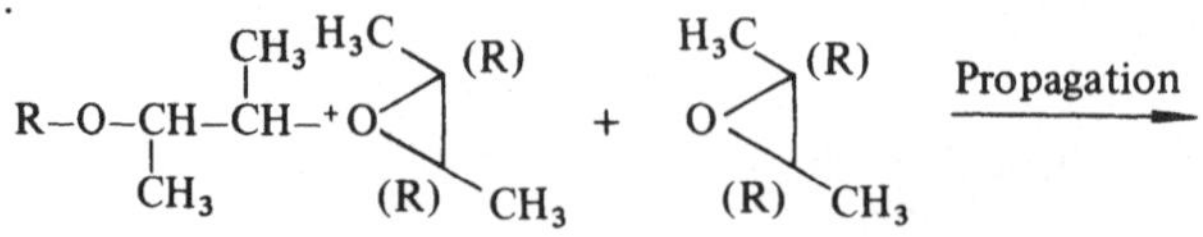

$$(87)$$

Attack on either of the equivalent carbon atoms in the oxonium ion leads to inversion of configuration and gives the required (R)(S) sequence. According to Vandenberg[147], molecular models of the transition state indeed indicate less steric hindrance to the attack of the enantiomer of the same stereochemistry[147]. Obviously, the reaction of the (S)(S) oxonium ion with the (S)(S) monomer leads by the same token to the (S)(R) sequence.

The polymerization of cis oxides gives, in contrast to trans monomers, polymers of different stereochemistry, namely cis-2,3-butene oxide forms racemic (RR, SS) disyndiotactic polymers whereas cis-1,4-dichloro-2,3-butene oxide gives racemic mixtures of (RR, RR) and (SS, SS) diisotactic polymers.

The formation of both kinds of polymers is explained on the basis of the interaction between the last polymer unit and the substituents in the oxonium ion:

$$
\begin{array}{c}
\text{(88)}
\end{array}
$$

88a

Disyndiotactic (RR, SS)
poly (cis−2,3−butene oxide)

88b

Diisotactic racemic (RR, RR) (SS, SS)
poly (cis−1,4−dichloro−2,3−butene oxide)

For Scheme (88a), according to the analysis of the stereochemical models, the methyl group of the last polymer unit may interact with the methyl group of the same stereochemistry within the oxonium ion ((S) in Scheme (88a)). This interaction facilitates the ring opening at the (S) carbon atom and eventually, due to the inversion of configuration, every (SS) sequence is followed by an (RR) sequence (and vice versa) giving the (SS, RR) disyndiotactic polymer.

The racemic diisotactic stereochemistry of poly(1,4-dichloro-2,3-butene oxide) requires a modification of the propagation mechanism proposed for 2,3-butene oxide. The dichloro monomer must attack the carbon atom in the oxonium ion with configuration opposite to that of the chain end. For example, in Scheme (88b) the (R) carbon atom must be attacked to form an (SS) unit which has the same configuration as the chain end. This leads to the poly (SS) polymer and, simultaneously, the poly(RR) chains will be formed. Vandenberg has proposed the interaction between the CH$_2$Cl group of the last unit (cf. Scheme (88b)) and the positively charged oxygen atom in order to remove this CH$_2$Cl group from the vicinity of the CH$_2$Cl group of similar stereochemistry in the oxonium ion. The same interaction has also been proposed by Vandenberg for the polymerization of cis-2,3-butene sulfide which also gives racemic mixtures of poly-(RR) and poly(SS) chains of a disotactic polymer.

Inversion of configuration in the chain growth was furthermore observed in the polymerization of 2-substituted 7-oxabicyclo[2.2.1]heptanes:

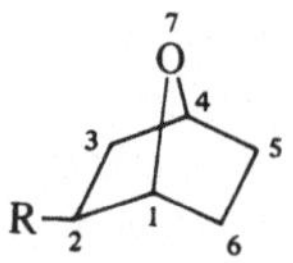

The polymerization of 2-endo- and 2-exo-methyl-7-oxabicyclo[2.2.1]heptanes[148]

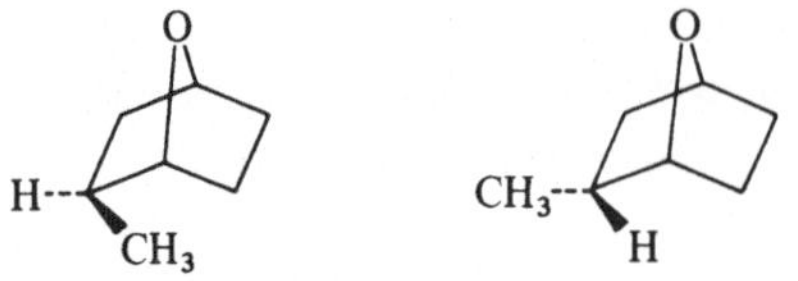

and the resulting polymers were studied by ^{1}H-NMR. The analysis of the spectrum of the polymer prepared from the endo monomer and the comparison with the spectrum of the model compound, trans-2-methyl-trans-4-hydroxycyclohexanol

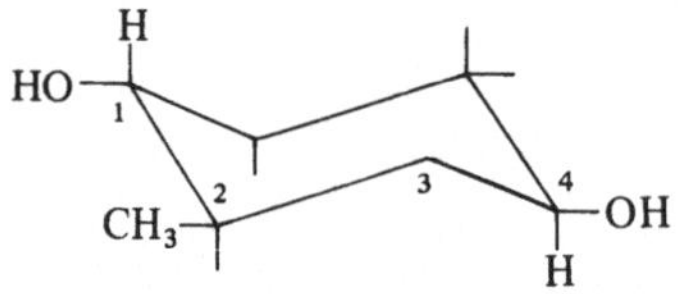

has allowed the structure of the polymer unit to be established[148]. This unit contains two equatorial ether groups and an equatorial methyl group:

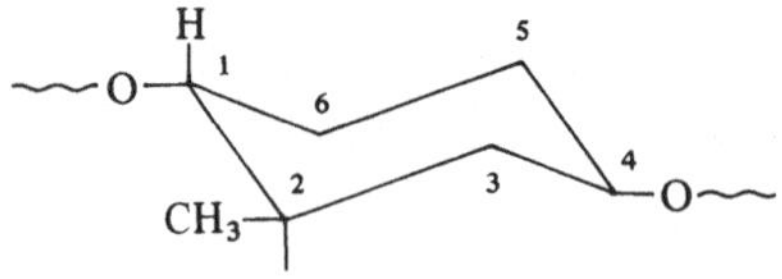

For the polymer of the exo monomer, the following structures have been established[148]:

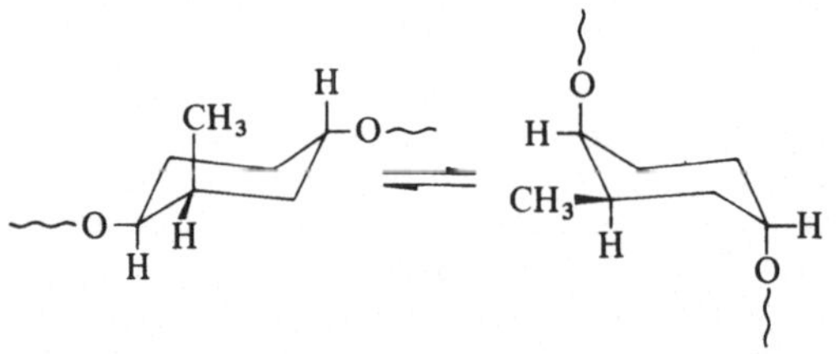

(89)

Thus, it has been shown that in the polymer of the endo monomer, the two ether groups as well as the methyl group are placed at the respective equatorial positions of the cyclohexane ring. This indicates inversion of configuration following an attack at C-4.

The same conclusion indicating inversion was reached from the structure of a polymer prepared from the exo monomer.

Kinetic studies revealed that the endo monomer propagates more rapidly, apparently because of its higher strain due to the interaction between the endo methyl group and the hydrogen atoms at C-3, C-5 and C-6[201]:

Similar interactions with the exo monomer should be considerably weaker[202].

Stereoregular polymers were also obtained by polymerizations of other cyclic monomers as shown by Scheurch[212], Hall[213] and Okada[214].

The majority of the stereochemical studies described in the above cited papers suggest an S_N2 mechanism of propagation. However, Hall observed in the polymerization of some bicyclic acetals (e.g. 2,6-dioxabicyclo[2.2.2]octane) the formation of a stereorandom polymer, indicating an S_N1 process. The randomization proceeds at higher temperatures whereas at low temperatures, a stereoregular polymer is exclusively formed by an S_N2 process[213].

However, the stereochemistry as a "diagnostic tool" to distinguish between S_N1 and S_N2 mechanisms has recently been questioned because racemization in the S_N1 process may not take place if a contact ion pair is involved and the anion sterically hinders the approach of the nucleophile from either side of the carbenium ion[147]. Since the agreement between the threo/erythro ratio in the polymer and the cis trans ratio in the monomer feed (e. g. 7:3 in both cases in the polymerization of ethylene oxide described above) is adequate, this uncertainty can be dismissed[146].

There is, however, still some doubt whether the polymerization of 2,2-dimethyloxirane (isobutylene oxide) also proceeds by S_N2 attack. This monomer is the best candidate for the S_N1 mechanism of propagation, because the generated dimethylcarbenium ion would be highly stabilized by the inductive effect:

$$\ldots-O-CH_2-\underset{\underset{CH_3}{|}}{\overset{\overset{CH_3}{|}}{C}}-{}^+O\diagup\hspace{-0.3em}\diagdown\overset{CH_3}{\underset{CH_3}{}} \rightleftharpoons \ldots-O-CH_2-\underset{\underset{CH_3}{|}}{\overset{\overset{CH_3}{|}}{C}}-O-CH_2-C{\overset{CH_3}{\underset{CH_3}{\diagdown}}}{}^+ \qquad (90)$$

Moreover, the acid hydrolysis of isobutylene oxide is believed to differ from that of other cyclic ethers studied. The following modified A_2 (A_2^+) mechanism has been proposed:

$$\underset{O}{\triangle}\overset{\overset{CH_3}{|}}{C}-CH_3 + H_3O^+ \rightleftharpoons \underset{\underset{H}{\overset{|}{O}}{}^{+/}}{\triangle}\overset{\overset{CH_3}{|}}{C}-CH_3 + H_2O \quad \text{(fast)} \quad (a)$$

$$\underset{\underset{H}{\overset{|}{O}}{}^{+/}}{\triangle}\overset{\overset{CH_3}{|}}{C}-CH_3 \rightleftharpoons HOCH_2-\underset{+}{\overset{\overset{CH_3}{|}}{C}}-CH_3 \quad \text{(fast)} \quad (b) \qquad (91)$$

$$HO-CH_2-\underset{+}{\overset{\overset{CH_3}{|}}{C}}-CH_3 + H_2O \longrightarrow HO-CH_2-\underset{\underset{+}{\overset{|}{O}H_2}}{\overset{\overset{CH_3}{|}}{C}}-CH_3$$

(c) (slow in comparison with reaction (b))

In this modification, the rate-determining step is again a bimolecular reaction involving, however, a carbenium ion.

In the usual S_N2 (A 2) mechanism, the reaction of the water molecule with the protonated compound is rate determining, while in the "pure" S_N1 (A 1) mechanism, the unimolecular opening of the protonated ring is rate determining. Discussing the mechanism of the polymerization of isobutylene oxide [147] it was eventually concluded that stereochemical studies would be needed to distinguish between these possibilities.

Oxiranes belong to the most strained heterocyclic monomers and to the weakest bases (only cyclic acetals are even weaker bases) which are discussed in this review.

Four-, five-, seven-, and higher-membered cyclic ethers, sulfides, and amines are all more nucleophilic and less strained that oxiranes. Thus, there are no grounds to expect that propagation of these monomers involves carbenium ions.

5 Termination and Transfer Processes

The initiation and propagation steps in the cationic polymerization of heterocycles have been described in previous sections. The chemistry of these steps has been reasonably well established, mostly by spectroscopic (particularly NMR) methods, and a number of corresponding rate constants have been determined.

Chain transfer and termination reactions leave their imprint primarily on the resulting structure of the end groups. Unless this structure is known any speculation about these two processes will be premature. The chemical nature of the end groups has, however, been studied only in few systems.

The borderline between transfer und termination is not very sharp in the literature and we shall use the following kinetic distinctions for transfer: it has no direct kinetic effect, growing species are fully restored, every act of transfer forms one "dead" macromolecule. We shall also discuss separately the special case of "temporary termination" being a reversible termination in which an active center becomes temporarily converted into its inactive (dormant) or much less active, isomeric counterpart. Termination forms one dead macromolecule and annihilates one active species.

5.1 Temporary (Reversible) Termination

5.1.1 Reversible Recombination with Non-Complex Anions

The growing macrocation can combine with an anion or with a fragment thereof and form a covalent bond. This reaction can either be reversible or irreversible and we shall confine ourselves in this section exclusively to the reversible ones.

In the following a few examples of reversible terminations through recombination of ion pairs are given:

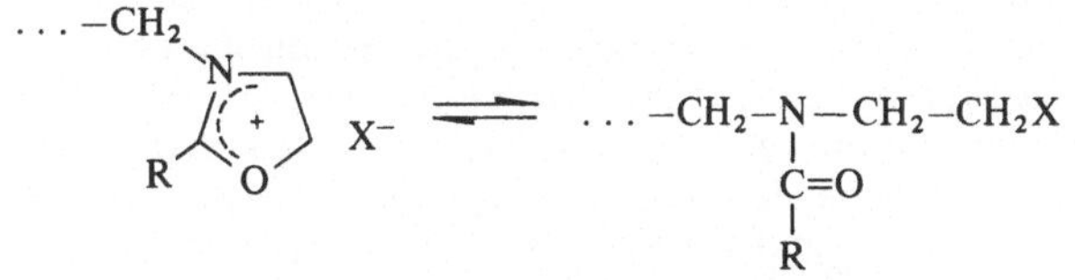

$$\ldots -CH_2 \underset{R}{\overset{}{\bigcirc}} X^- \; \rightleftharpoons \; \ldots -CH_2-N-CH_2-CH_2X \atop \underset{R}{\overset{|}{C}=O} \tag{92}$$

(where X = e.g. halogen[99], ClO_4[216])

The equilibrium according to Eq. (92) involving oxazolinium cations was apparently the first observation of the equilibrium between ionic and covalent species in the cationic polymerization of heterocyclic monomers[207]. A similar collapse, being however irreversible, has earlier been postulated for the cationic polymerization of oxetane propagating with $^-C(NO_2)_3$ anions[217] and of styrene propagating with ClO_4^- anions[218].

In the polymerization of THF, the kinetics and thermodynamics of the equilibrium between covalent macromolecules (macroesters) and macroion pairs have recently been elucidated: e.g.; for $CF_3SO_3^-$ anion:

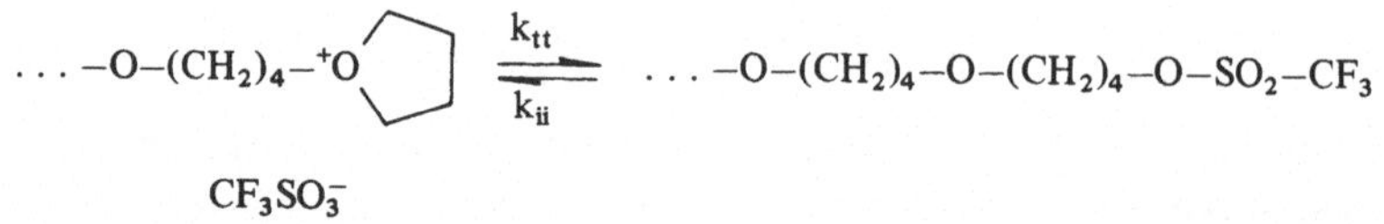

$$\ldots-O-(CH_2)_4-{}^+O \overset{k_{tt}}{\underset{k_{ii}}{\rightleftarrows}} \quad \ldots-O-(CH_2)_4-O-(CH_2)_4-O-SO_2-CF_3 \qquad (93)$$

$$CF_3SO_3^-$$

This equilibrium was postulated by Smith several years ago[100]; in the same year it was strongly supported by kinetic evidence[219] and, although the macroester was not detected by ^{1}H-NMR[220] at the beginning, this equilibrium was later amply documented by ^{1}H-, ^{19}F-, and ^{13}C-NMR spectra[138, 126, 221, 222].

The use of high-resolution ^{1}H-NMR (300 MHz) was the most useful method of solving the early confusion[138].

First, it is instructive to examine the actual ^{1}H-NMR spectra of the living poly-THF with $CF_3SO_3^-$ anions in three different solvents, namely CCl_4, CH_2Cl_2 and CH_3NO_2 (sometimes perdeuterated solvents are used to facilitate the observation of the region of the growing species).

In the ^{1}H-NMR (300 MHz) spectrum (Fig. 9), the actual chemical shifts may slightly depend on concentration, solvent and anion structure or temperature, but these differences for a given species (ion or ester) are small as compared with the differences between ions and esters.

The corresponding triplets observed in the spectra (Fig. 9) were identified by comparing their chemical shifts with those of living poly-THF containing SbF_6^- anion (only one triplet characteristic of the ion) and with the low molecular esters of triflic acid.

Thus, the following chemical shifts were attributed to these two isomeric forms, e. g. for the FSO_3^- anion:

$$(94)$$

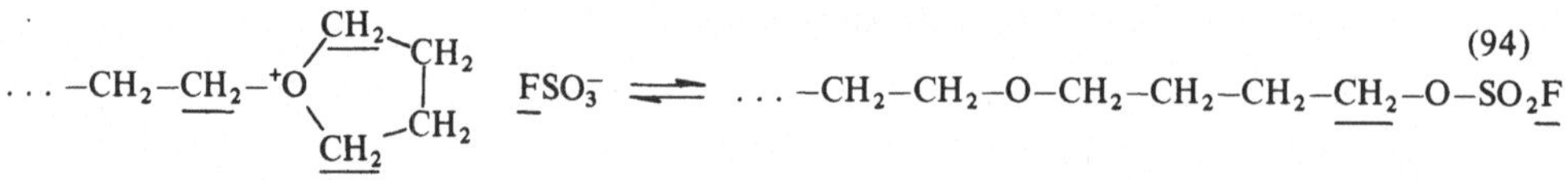

$$\ldots-CH_2-CH_2-{}^+O\overset{CH_2-CH_2}{\underset{CH_2-CH_2}{|}} \quad FSO_3^- \rightleftharpoons \ldots-CH_2-CH_2-O-CH_2-CH_2-CH_2-CH_2-O-SO_2F$$

^{1}H –NMR:	4.89	4.85		4.56[13]
^{19}F –NMR:		34.5		38.0[221]
^{13}C –NMR:	89.6	87.5		79.1[222]

Another system for which the simultaneous presence of the covalent and ionic species was unambiguously shown is the polymerization of oxazolines. Although the participation of the covalent species was proposed in an early series of works of Kagiya on substituted oxazolines[216, 223], the chemistry of polymerization of these monomers was not clearly defined, and various cannonical forms of ions were erroneously treated as thermodynamically distinguishable species. This was rectified a few years later by Saegusa's group working with different oxazolines, mostly 2-oxazoline, 2-methyl- and 2-phenyl-2-oxazolines[99, 207].

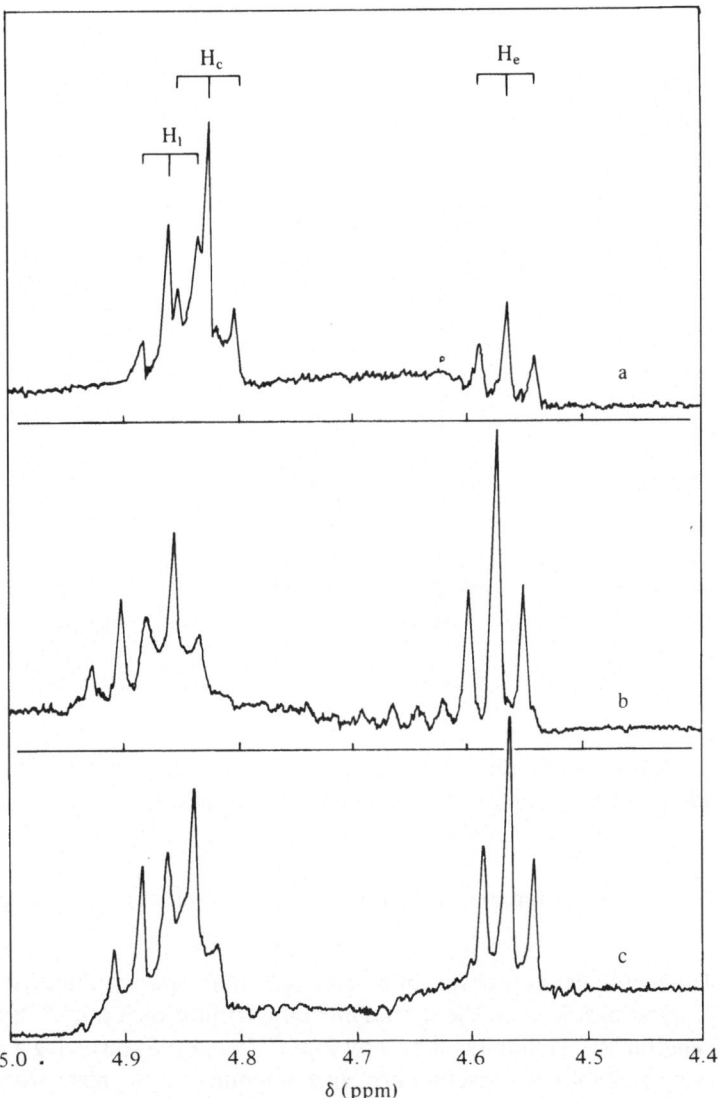

Fig. 9a–c. ^{1}H-NMR (300 MHz) spectra of the region of active species in the polymerization of THF at $-18\,°C$ with $CF_3SO_3^-$ anion in CH_2Cl_2/CCl_4 (a), CCl_4 (b) and at $+18\,°C$ in CH_2Cl_2 (c). $[THF]_0 = 8.0$ mole $\cdot$ l^{-1}, $[I]_0 = 10^{-1}$ mole $\cdot$ l^{-1}

2,3-Dimethyl-2-oxazolinium bromide was prepared as a model compound of the active species and its solution was studied by ^{1}H-NMR:

$$
\begin{array}{c}
\overset{(1)}{H_3C}\diagdown \\
\qquad N\!-\!\overset{(4)}{CH_2} \\
\overset{(2)}{H_3C}\!-\!C\diagup\ \diagup_{(+)}\ \diagdown_{(3)}\quad ,\ Br^- \rightleftharpoons \overset{(5)}{H_3C}\!-\!N\!-\!\overset{(7)}{CH_2}\!-\!\overset{(7)}{CH_2}Br \\
\qquad\qquad O \qquad\qquad\qquad\qquad |\!\!\!\!\!\quad\ C \\
\qquad\qquad\qquad\qquad\qquad\qquad\quad\ \overset{(6)}{H_3C}\diagup\ \diagdown O
\end{array}
\qquad (95)
$$

The corresponding ^{1}H-NMR spectrum in CD_3CN at 35 °C is described in Fig. 10.

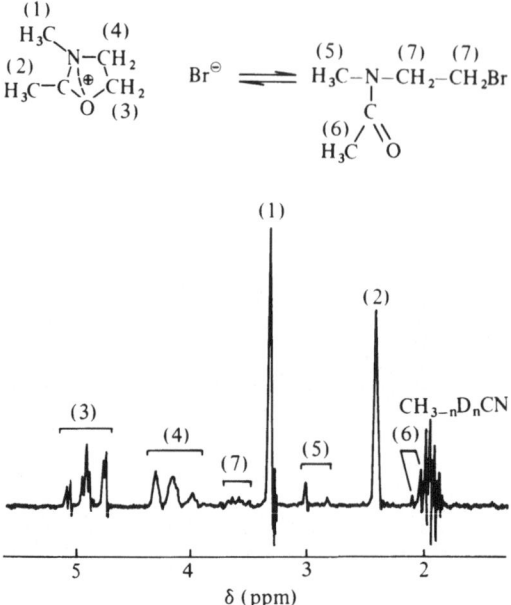

Fig. 10. ^{1}H-NMR spectrum of the system (95) in CD_3CN at 35 °C (Ref. 99)

When a Cl^- anion was used in the salt, only the covalent form was observed while for the J^- anion, only the ionic form appeared[99].

5.1.2 Competition Between Propagation of Covalent and Ionic Species

The simultaneous presence of covalent and ionic species immediately poses a number of questions: how do the species interconvert into each other? By a unimolecular penultimate reaction (as shown by Eqs. (94) and (95) involving a polymer segment or by a bimolecular reaction involving monomer or polymer? How does this equilibrium depend on conditions (solvent, temperature)? How important is the contribution of both kinds of species to propagation? Let us examine these questions starting from the last one and taking the polymerization of THF as an example.

When two kinds of species propagate simultaneously and their proportions are known from independent measurements, then the rate constants of propagation on both kinds of competing (or cooperating?) species can be determined from a simple relationship:

$$k_p^{app} = \beta\, k_p^i + (1 - \beta)\, k_p^c \tag{96}$$

where k_p^{app} is the apparent second-order rate constant of propagation observed directly in experiment, k_p^i the rate constant of propagation of ionic species, k_p^c the rate constant of propagation of covalent species, and β the proportion of macroions in the system:

$$\cdot(\ldots)_n{-}CH_2{-}{}^+O\overset{\frown}{\underset{}{\bigcirc}}\ ,A^- \ +\ O\overset{\frown}{\underset{}{\bigcirc}}\ \underset{k_d}{\overset{k_p^i}{\rightleftharpoons}}\ \cdot(\ldots)_{n+1}{-}CH_2{-}{}^+O\overset{\frown}{\underset{}{\bigcirc}}$$
$$A^- \qquad\qquad\qquad\qquad\qquad\qquad\qquad\qquad\qquad A^-$$

$$\cdot(\ldots)_n{-}CH_2{-}O{-}CH_2{-}CH_2{-}CH_2{-}CH_2{-}A\ +\ O\overset{\frown}{\underset{}{\bigcirc}}\ \underset{k_{tt}'}{\overset{k_p^c}{\rightleftharpoons}} \tag{97}$$

$$\cdot(\ldots)_n{-}CH_2{-}O{-}CH_2{-}CH_2{-}CH_2{-}\ CH_2{-}{}^+O\overset{\frown}{\underset{}{\bigcirc}} \tag{98}$$
$$A^-$$

Eq. (96) in the form:

$$k_p^{app}/\beta - k_p^i = -k_p^c + k_p^c/\beta \tag{99}$$

should yield k_p^c by simply plotting $k_p^{app}/\beta - k_p^i$ as a function of β^{-1}, provided that k_p^i in the polymerization of THF is the same for various anions and can be determined independently. In order to change β, either the polarity of the medium or the overall concentration of the growing species can be changed. Simple equilibria like (94) and (95) describe reality only under the condition that the other two processes, which might involve ion pairs, do not participate at the same time. These are: dissociation of ion pairs into ions and aggregation of ion pairs to higher aggregates. Studies taking into account these complications have hitherto been performed only for one system, namely for the polymerization of THF with $CF_3SO_3^-$ anion[224].

The plot of Eq. (99) is shown in Fig. 11 by a solid line 1a.

Unfortunately, the slope is too small to give a reliable value of k_p^c and the intercept does not sufficiently differ from zero. In order, however, to at least estimate

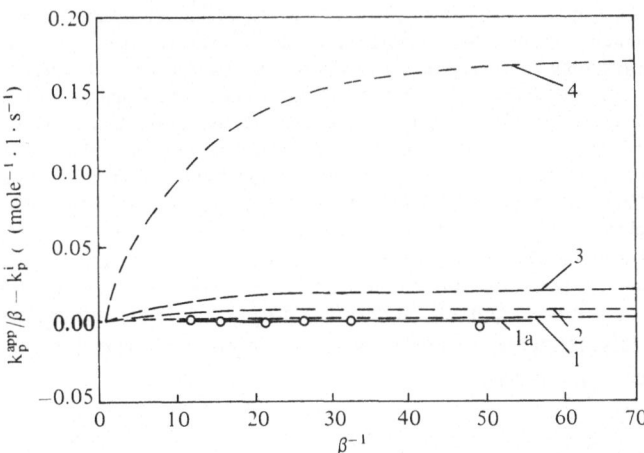

Fig. 11. Determination of k_p^i and k_p^C (ionic and covalent propagations) in the polymerization of THF. Plot of $k_p^{app} \cdot \beta^{-1} - k_p^i$ as a function of β^{-1} for the polymerization of THF with $CF_3SO_3^-$ anion at 25 °C. Line 1a (solid): experimental points. Broken lines 1,2,3,4 are calculated taking k_p^c arbitrarly equal to k_a, $5 \cdot k_a$, $10 \cdot k_a$, and $10^2 \cdot k_a$, respectively (Ref. 224)

the value of k_p^c the following procedure is adopted[224]: it is first assumed that $k_p^c = k_a$ where k_a is a rate constant of the reaction between ethyl triflate (a low-molecular weight model of a macroester) and THF:

$$C_2H_5OSO_2CF_3 \; + \; O\!\!\big\langle\!\!\big\rangle \; \overset{k_a}{\rightleftharpoons} \; C_2H_5\overset{+}{-}O\!\!\big\langle\!\!\big\rangle \;, \; {}^-OSO_2CF_3 \tag{100}$$

The values of k_a are then determined separately (cf. Sect. III) and when the appropriate values of k_a and k_p^i (for a given β value since both constants depend on the polarity of the medium) are introduced into Eq. (99), the broken line 1 is obtained. It coincides quite well with the experimental solid line 1. This agreement between two lines simply means that $k_p^c \lessapprox k_a$, i.e. that the rate constant of the addition of THF to the macroester (macrotriflate) is not larger than the rate constant of the addition of THF to ethyl triflate. Indeed, if larger values of k_p^c are inserted to Eq. (99) (e.g. $k_p^c = 5 \cdot k_a$, $10 \cdot k_a$ and $10^2 \cdot k_a$, these values corresponding to broken lines 2, 3 and 4 in Fig. 11), then the experimental points of line 1 cannot be fitted any more to these calculated lines.

Thus, for the polymerization of THF in all solvents studied we have $k_p^i \gg k_p^c$. However, the ratio of these two constants depends on the bulk polarity of the system. This is because k_p^i decreases and k_p^c increases with polarity (cf. Sect. 3.3 and 4.3). The largest difference between k_p^i and k_p^c is observed in CCl_4 (approx. 10^3 times) and the lowest in CH_3NO_2 (approx. 50 times).

Another group working on the same problem has come to different conclusions, namely that in the reaction with THF the macroester is much more reactive than low-molecular esters like ethyl triflate[225].

The main difference between the experimental conditions of these two groups involves the concentration of the active species being equal or below 10^{-1} mole $\cdot$ 1^{-1} in the work of the former group and as large as 1.0 mole $\cdot$ 1^{-1} (for e.g. $[THF]_0 = 5.0$ mole $\cdot$ 1^{-1}) in the latter.

The large concentrations of initiator may lead to the precipitation of a portion of the macroions; this was indeed observed when polymerization performed in CCl_4 was rechecked[88]. The precipitation of the low molecular macroions has also recently been reported for dicationic species[101]. The precipitated macroions become undetectable by NMR but can still be reactive in propagation. Obviously, this will result in the wrong conclusion that all of the observed growth is due to macroesters and will give a value for k_p^c larger than the real one.

The polymerization of 2-methyl-2-oxazoline is another example of the parallel growth of macroions and covalent macromolecules[99]. The averaged rate coefficients of propagation $\overline{k}_p$ (including k_p^i and k_p^c) for various anions, namely Cl^-, Br^-, J^-, and ${}^-OTos$ (Tos $= SO_2C_6H_4CH_3$-p) were measured in CD_3CN at 40 °C and found to be equal to: ($\times 10^4$, in mole$^{-1} \cdot 1 \cdot s^{-1}$) 0.03, 1.2, 1.14 and 1.17 (listed in the same order as the anions). Thus, k_p is almost the same for Br^-, J^-, and ${}^-OTos$ anions and it was assumed that these k_p values are equal to the propagation rate constant of ionic species. Since the propagation rate constant with Cl^- anions equals $0.03 \cdot 10^{-4}$ mole$^{-1} \cdot 1 \cdot s^{-1}$, this was attributed to covalent growth. ^{1}H-NMR spectra for Cl^- exclusively show covalent species whereas for Br^-, J^- and ${}^-OTs$ anions only ionic species could be found. Thus, both kinetic and NMR data are in good agreement and this system seems to be eminently suitable for further detailed studies of ionic vs. covalent reactivities.

Thus, in general, in cationic polymerization, and particularly in the cationic polymerization of heterocyclic monomers, propagation can involve both ionic and covalent species:

$$\mathrm{-\!\!\!(\ldots)_{\overline{n}}\!- A + M \underset{k'_{tt}}{\overset{k^c_p}{\rightleftharpoons}} -\!\!\!(\ldots)_{\overline{n}}\!- \; ^+M, A^-}$$

$$K_e = k_{ii}/k_{tt}$$

$$k_{ii} \Big\uparrow\Big\downarrow k_{tt} \tag{101}$$

$$\mathrm{-\!\!\!(\ldots)_{\overline{n-1}}\!- {}^+M, A^- + M \underset{k_d}{\overset{k^i_p}{\rightleftharpoons}} -\!\!\!(\ldots)_{\overline{n}}\!- {}^+M, A^-};$$

$$\frac{k'_{tt}}{k_{tt}} = \frac{k_d}{k^i_p} \cdot \frac{k^c_p}{k_{ii}}$$

The proportion of covalent macromolecules $\mathrm{-\!\!(\ldots)_{\overline{n}}\!- A}$ and their ionic counterparts $\mathrm{-\!\!(\ldots)_{\overline{n}}\!- {}^+M, A^-}$ in the chain growth depends on the relative concentrations of both species, i.e. on K_e and on the ratio k^c_p/k^i_p (the ratio k'_{tt}/k_d being obviously equal to that of $k^c_p/k^i_p \cdot K_e$ as demanded by a principle of reversibility of microstates).

It is of interest to note that the systems studied till now are "either-or", i.e. involving propagation exclusively dominated either by ionic species or by covalent ones. Up to now, the ionic species were found to be much more reactive than covalent ones. This can be explained on the basis of two models of S_N2 reactions involving attack on covalent and ionic species, e.g.:

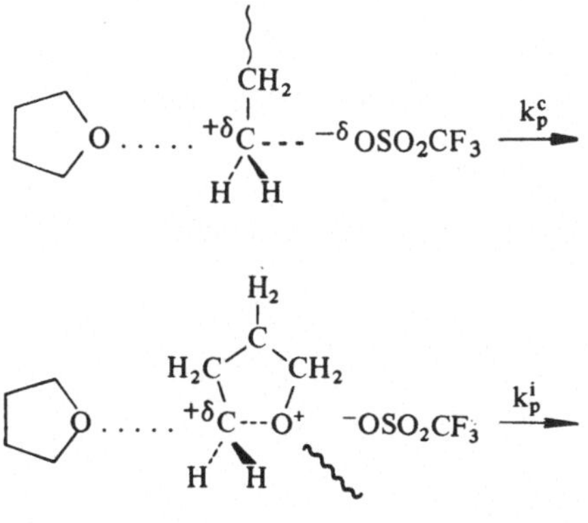

$$\tag{102}$$

The rupture of the $\overset{+\delta}{\underset{\bigwedge}{\mathrm{C}}} - \mathrm{O}^{-\delta}-$ bond in the covalent species requires more energy than that of the $\overset{+\delta}{\underset{\bigwedge}{\mathrm{C}}} - \overset{+}{\underset{|}{\mathrm{O}}}-$ bond in the oxonium ion because the strain is partially released in the transition state of the latter structure. On the other hand, since covalent species like macroesters are not as strongly solvated as the ions, solvation of the more ionic transition state is an additional driving force for covalent propagation. Thus, with increasing solvating power of the solvent used, the k^c_p values also increases. Since, as already mentioned, k^i_p decreases with increasing solvent polarity, one could expect that under certain conditions k^c_p would become even larger than k^i_p. Simple calculations, based on the determined dependences of k^i_p and k^c_p on polarity expressed by the dielectric constant ϵ, in the polymerization of THF[226] reveal that even for the dielectric constant equal to infinity $k^i_p > k^c_p$, provided that the observed dependences are obeyed up to this unrealistic ϵ value.

In other systems not yet described, this inversion of reactivities may, however, occur in an accessible range of polarities.

5.1.3 Equilibria Between Covalent and Ionic Growing Species

In the polymerization of cyclic ethers, the equilibria of a number of systems have been studied. However, oxirane and oxetane give exclusively macroesters[130, 131] apparently because the conversion of the macroester into the macroion, which is (as it will be shown in this section) a unimolecular reaction within a last polymer unit, is energetically unfavorable:

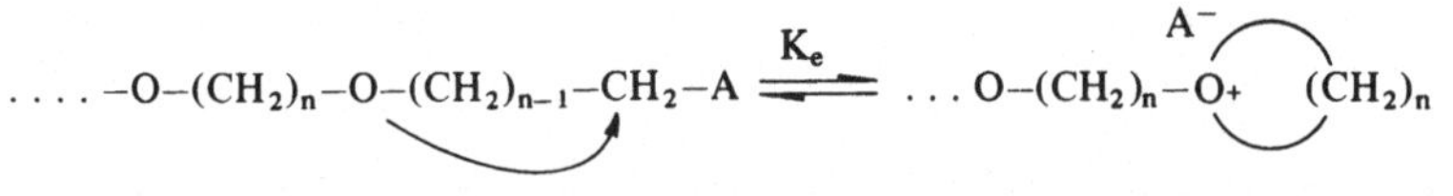

$$\ldots\ldots -O-(CH_2)_n-O-(CH_2)_{n-1}-CH_2-A \xrightleftharpoons{K_e} \ldots O-(CH_2)_n-\overset{+}{O} \overset{A^-}{\underset{\frown}{(CH_2)_n}} \tag{103}$$

For n = 2 or 3, a large ring strain (approx. 20 kcal $\cdot$ mole^{-1} in monomers) has to be overcome in order to transform a macroester into a three — or four-membered oxonium ion. When n = 2, however, ionic species are formed in which the two last units are engaged. Thus, the attack involves the penultimate unit:

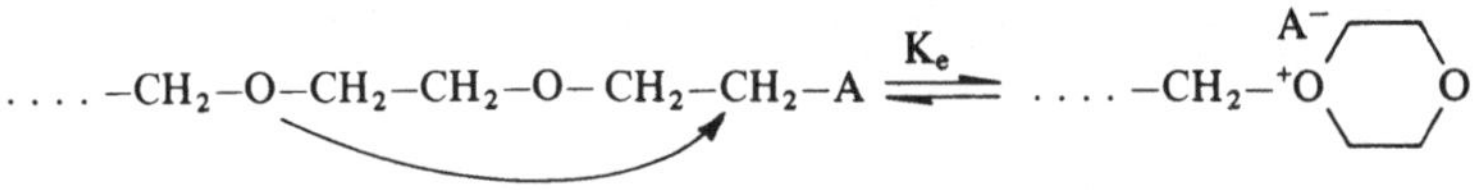

$$\ldots\ldots -CH_2-O-CH_2-CH_2-O-CH_2-CH_2-A \xrightleftharpoons{K_e} \ldots\ldots -CH_2-\overset{+}{O} \overset{A^-}{\bigcirc} O \tag{104}$$

Since the six-membered ring is formed, almost no ring strain needs to be overcome.

In similar equilibria, involving nitrogen-containing compounds, the 6-membered ring was found to be formed 10 times more rapidly than the three-membered one[205].

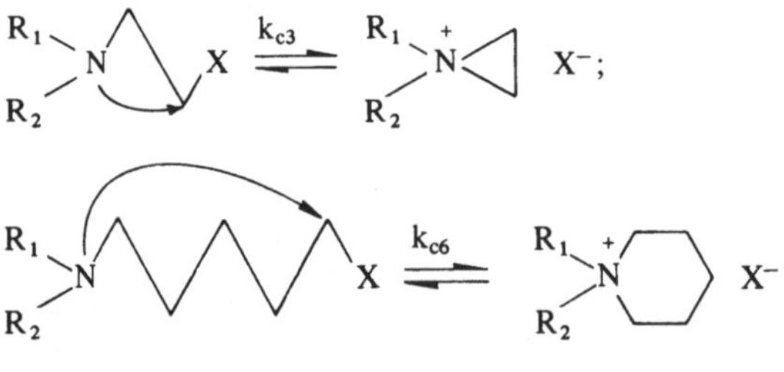

$$\begin{array}{c} R_1 \\ \diagdown \\ N \\ \diagup \\ R_2 \end{array} \overset{\triangle}{} X \xrightleftharpoons{k_{c3}} \begin{array}{c} R_1 \\ \diagdown \overset{+}{N}\triangleleft \\ \diagup \\ R_2 \end{array} X^-;$$

$$\tag{105}$$

$$\begin{array}{c} R_1 \\ \diagdown \\ N \\ \diagup \\ R_2 \end{array} X \xrightleftharpoons{k_{c6}} \begin{array}{c} R_1 \\ \diagdown \overset{+}{N} \\ \diagup \\ R_2 \end{array} X^-$$

The thermodynamics of equilibrium of the polymerization of THF

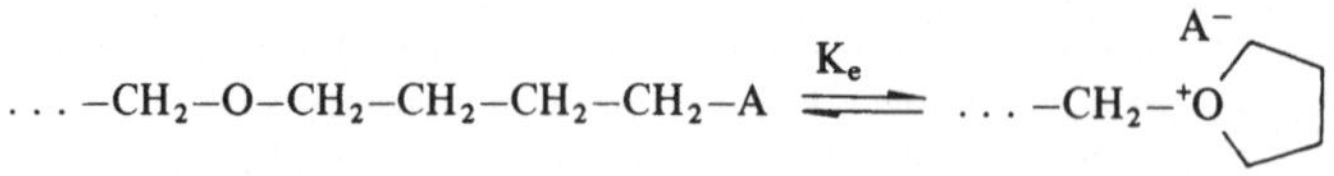

$$\ldots -CH_2-O-CH_2-CH_2-CH_2-CH_2-A \xrightleftharpoons{K_e} \ldots -CH_2-\overset{+}{O} \overset{A^-}{\bigcirc} \tag{106}$$

was studied by two groups but only one group succeeded in securing the necessary conditions whereby the equilibria involved were exactly described by Eq. (106). Indeed, the more complete equilibrium, which cannot quantitatively be treated by a simple determination of the ratio [ions]/[covalent species], involves two more reactions, namely dissociation of the ion pairs into ions and/or aggregation of the ion pairs into higher aggregates of ion pairs:

$$\ldots -CH_2-O-CH_2-CH_2-CH_2-CH_2-A \underset{k_{tt}^{\pm}}{\overset{k_{ii}^{\pm}}{\rightleftharpoons}} \ldots -CH_2-{}^{+}O \bigcirc$$

$$\text{Aggregation (pairs of pairs etc.)}$$

$$A^{-} \quad K_D \qquad (107)$$

$$\ldots -CH_2-{}^{+}O \bigcirc$$

$$+ A^{-}$$

$$K_e^{\pm} = k_{ii}^{\pm}/k_{tt}^{\pm}$$

At higher total concentration of growing species (Σ [macroions] + [macroesters]), the proportion of ions increases[228]. This occurs in the region of concentrations exceeding 0.1 mole $\cdot$ l^{-1} where practically macroion pairs are the only ionic species present (free ions are absent), aggregating at these higher concentrations and thus shifting the equilibrium between macroesters and macroion pairs to the direction of macroions, because of the formation of aggregates.

There is no way at present to measure directly the concentration of aggregated macroion pairs separately from the non-aggregated ones. Thus, in order to determine the thermodynamic equilibrium constant K_e, it is necessary to use the experimental conditions excluding the formation of aggregates. It has been experimentally verified that the proportion of macroions becomes independent (in CH_2Cl_2) of the total concentration of active species only below 0.1 mole $\cdot$ l^{-1}. Thus, only the results obtained at this concentration or below can be used in the computation of K_e according to the simple relationship: K_e = [macroion pairs]/[macroesters]. Thus, K_e obtained in this way and the corresponding thermodynamic parameters are listed in Table 13.

Table 13. Dependence of K_e (= k_{ii}/k_{tt}), ΔH_e^0 and ΔS_e^0 on the polarity of the medium for $CF_3SO_3^-$ anion [229]

Solvent	[THF] (mole $\cdot$ l^{-1})	Dielectric constant of the solvent – monomer mixture	K_e 25 °C	ΔH_e^0 (kcal $\cdot$ mole^{-1})	ΔS_e^0 cal $\cdot$ mole$^{-1}\cdot$K^{-1}
CCl$_4$	8.0	5.9	0.06	-6.0 ± 0.3	-25 ± 1
CH$_2$Cl$_2$	8.0	8.2	0.58	-5.6 ± 0.2	-19.7 ± 0.8
CH$_3$NO$_2$	8.0	20.6	42.0	–	–

There is an analogy in the change of the thermodynamic functions in the reversible ionization described above and the better known dissociation of ion pairs into free ions. In both processes, solvation factors prevail and are more important than the internal rearrangement of the species involved. Thus, the exothermicity of ionization may partially originate in ionization itself (attack of the oxygen atom on the carbon atom of the polarized ester group) and in the heat of solvation of the more ionic macroion pair. The negative entropy of ionization is due to the immobilization of monomer and solvent molecules in the proximity of the ion pairs. Interaction with ions is stronger than with a macroester; thus, monomer and solvent molecules are better organized around ions[229].

Similar conclusions were formulated on the basis of measurements performed for FSO_3^- and ClO_4^- anions in CH_3NO_2. The equilibrium constants K_e were found to decrease in the following order (K_e measured at 20 °C for $[THF]_0 = 8.0$ mole $\cdot$ l^{-1}): $CF_3SO_3^-$ (42.0), FSO_3^- (20.0), ClO_4^- (9.5). Thus, the higher the nucleophilicity of anion, the lower K_e, due to the increased probability of collapse of the ion pair[230].

In similar studies of the polymerization of 2-methyl-2-oxazoline initiated with methyl bromide, ionic 2,3-dimethyloxazolinium bromide is in equilibrium with its covalent counterpart[99]:

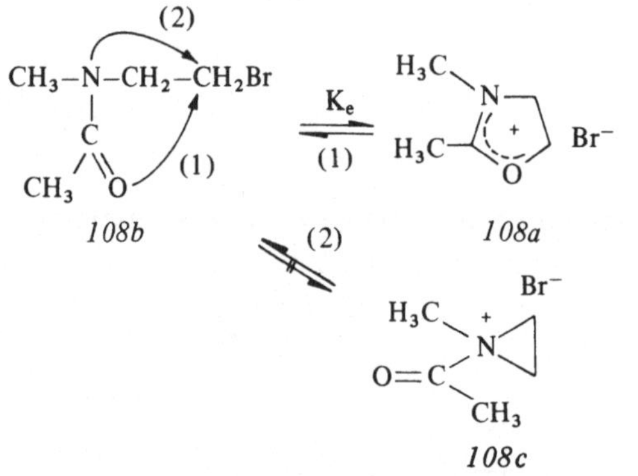

(108)

Although the nitrogen atom in *108b* is much more nucleophilic than the carbonyl oxygen atom, the aziridinium cation *108c* is not observed by NMR. Actually, in the reported ^{1}H-NMR spectrum[99], the signals due to methylene groups appear as a multiplet resembling a poorly resolved AB pattern which is in better agreement with *108a* than with *108c*. The formation of *108a* and not of *108c* is due to the large ring strain accompanying the closure of the 3-membered ring.

Equilibrium measurements according to Eq. (108) were performed at quite high concentrations of active species ($\sim$0.6 mole $\cdot$ l^{-1}) and it is not certain whether these measurements actually gave the equilibrium constants (and related thermodynamic parameters) or whether the determined values are the composite coefficients involving two or more equilibria.

Table 14. Dependence of K_e, ΔH_e and ΔS_e on the polarity of the medium for an equilibrium

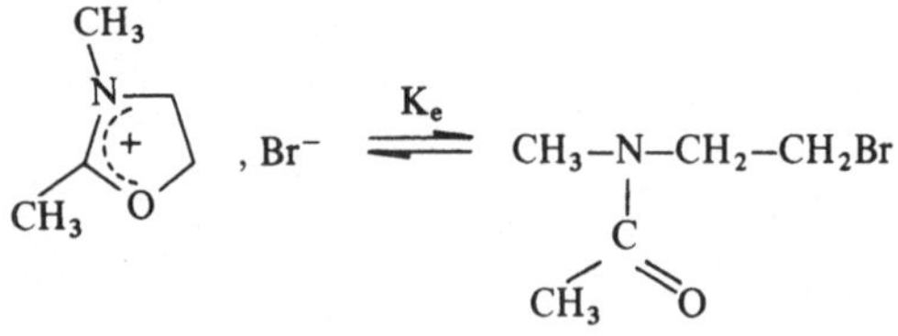

modelling macroion pair – covalent macromolecule interchange in the polymerization of 2-methyl-2-oxazoline[99]

Solvent	Dielectric constant	K_e (35 °C)	ΔH_e (kcal $\cdot$ mole^{-1})	ΔS_e (cal $\cdot$ mole$^{-1} \cdot$ K^{-1})
$CHCl_3$	4.7	4.0	-7.4	-21
CD_3CN	34.0	10.0	-5.5	-13
CD_3NO_2	35.0	14.3	-4.1	-8.1

Assuming tentatively that the determined values have a thermodynamic meaning, the observed trends can be compared with those found for THF.

Thus, both systems give a qualitative agreement: ionization is exothermic and accompanied by a decrease of entropy. In more polar solvents, the exothermicity decreases. ΔH_e becomes less exothermic in more polar solvents; this means that the enthalpy of the ground state of covalent species is less affected than that of the ionic species. The graphical explanation is shown below in Fig. 12.

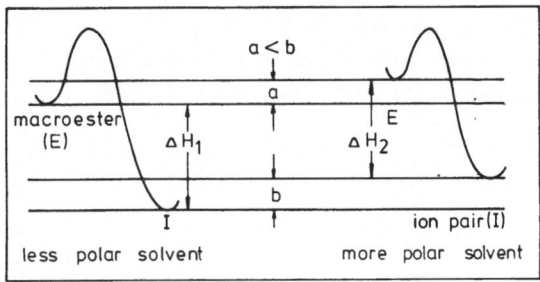

Fig. 12. Diagram of enthalpy change in the macroester-macroion pair interconversion

5.1.4 Equilibrium Between Covalent Macromolecules and Macroion Pairs in the Presence of Macroions

In polar solvents with high ionizing power and at lower concentrations of active species, macroion pairs dissociate into ions, shifting the equilibrium and increasing in this way the overall proportion of ions:

$$\ldots -CH_2-O-CH_2-CH_2-CH_2-CH_2-OClO_3 \overset{K_e}{\rightleftharpoons} \ldots -CH_2-{}^+O\!\!\bigcirc\, , ClO_4^- \overset{K_D}{\rightleftharpoons}$$

$$\rightleftharpoons \ldots -CH_2-{}^+O\!\!\bigcirc\; + \; ClO_4^- \tag{109}$$

In this system the simple ratio of [ions]/[macroester], which can be directly measured, does not give K_e. In order to determine K_e one has to use the following equation[230]:

$$\frac{[P^+]+[P^\pm]}{[I]_0} = \frac{1 + 2/K_e + \sqrt{1 + 4[I]_0\,(1+K_e)/K_e \cdot K_D}}{1 + \sqrt{1 + 4[I]_0\,(1+K_e)/K_e \cdot K_D}} \cdot \left(\frac{K_e}{1+K_e}\right) \tag{110}$$

If it is assumed that K_D (determined independently) does not change in the presence of macroesters, then K_e can be found numerically by using Eq. (110).

5.1.5 The Rate of Interconversion of Covalent Macromolecules and Macroion Pairs

The rate of interconversion was measured for one monomer only, namely in the polymerization of THF with FSO_3^- and $CF_3SO_3^-$ anions. It was determined by using the "temperature jump" technique for the equilibrated systems. Thus, a system was brought to the macroions – macroester equilibrium at the given temperature, then the temperature was suddenly changed and the approach to the new equilibrium at this temperature was observed by monitoring (by ^{1}H-NMR–300 MHz) the concentration of macroesters. The typical ^{1}H-NMR spectra, showing the change of the concentrations with time, are given in Fig. 13.

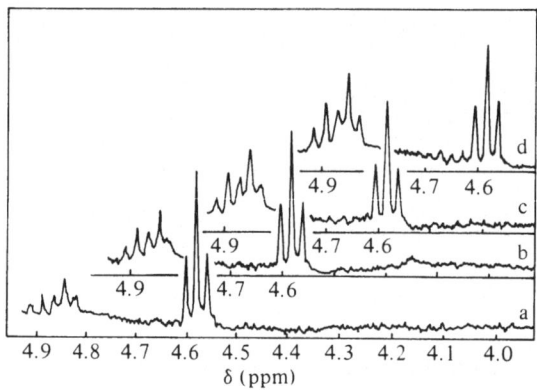

Fig. 13a–d. Determination of k_{ii} and k_{tt} by "temperature jump". Dependence of ^{1}H-NMR (300 MHz) spectra of living ends on time in the polymerization of THF with $CF_3SO_3^-$ anion at $-18\,°C$ after disrupting an equilibrium established at $+18\,°C$ (after time s x 10^2) : (**a**) 1.5, (**b**) 5.7, (**c**) 13.2, (**d**) 22.8. CCl_4 solvent, $[THF]_0 = 8.0$ mole $\cdot$ l^{-1}, $[CF_3SO_3CH_3]_0 = = 8 \cdot 10^{-2}$ mole $\cdot$ l^{-1} (Ref. 229)

As can be seen from Fig. 13, the systematic decrease of the macroester triplet at $\delta = 4.58$ ppm is accompanied by an equivalent increase of the macroion pair triplets.

The macroester $\rightleftharpoons$ macroion pair interconversion has been analyzed as an unimolecular opposed reaction:

$$\ldots -CH_2-O-(CH_2)_4-O-SO_2-CF_3 \underset{k_{tt}^\pm}{\overset{k_{ii}^\pm}{\rightleftharpoons}} \ldots -CH_2-{}^+O\!\!\bigcirc,\ {}^\ominus O-SO_2-CF_3 \qquad (111)$$

$$E \phantom{-SO_2-CF_3 \underset{k_{tt}^\pm}{\overset{k_{ii}^\pm}{\rightleftharpoons}} \ldots -CH_2-{}^+O\!\!\bigcirc}P^\pm$$

The rate constant of internal ionization $k_{ii}^\pm$ and the rate constant of temporary termination $k_{tt}^\pm$ (collapse of the ion pair) were determined from the corresponding plots of Eq. (112):

$$\ln \left(\frac{[E]_0 - [E]_e}{[E]_t - [E]_e} \right) \Bigg/ \frac{(1 + K_e^\pm)}{K_e^\pm} = k_{ii}^\pm \cdot t \ ; \quad K_e^\pm = \frac{k_{ii}^\pm}{k_{tt}^\pm} \qquad (112)$$

Table 15. Rate constants and activation parameters of internal ionization (ii) and temporary termination (tt) in the polymerization of THF with triflic ($CF_3SO_3^-$) and fluorosulfonic (FSO_3^-) anions ($[THF]_0 = 8$ mole $\cdot$ l^{-1}, 25 °C) (Ref. 229)

Solvent	Anion	$k_{ii} \cdot 10^2$ (s^{-1})	$\Delta H_{ii}^{\ddagger}$ (kcal $\cdot$ mole^{-1})	$\Delta S_{ii}^{\ddagger}$ (cal $\cdot$ mole$^{-1} \cdot$ K^{-1})	$k_{tt} \cdot 10^2$ (s^{-1})	$\Delta H_{tt}^{\ddagger}$ (kcal $\cdot$ mole^{-1})	$\Delta S_{tt}^{\ddagger}$ (cal $\cdot$ mole$^{-1} \cdot$ K^{-1})
CCl_4	$CF_3SO_3^-$	0.8	10.4 ± 1.2	-33 ± 5	12.1	16.4 ± 1.5	-8 ± 6
CH_2Cl_2	$CF_3SO_3^-$	1.9	8.2 ± 1.0	-39 ± 4	3.3	13.8 ± 1.2	-19 ± 5
CH_2Cl_2	FSO_3^-	1.6	11.3	-29	5.9	16.0	-10

The rate constant $k_{ii}^{\pm}$ has to be distinguished from the rate constant of external ionization, k_{ei}, called in this review rate constant of propagation of the macroester (see p. 81) (thus, the meanings of k_{ei} and k_p^c are the same) and related to reaction (113):

$$... -O\!\!-\!\!(CH_2)_{\overline{4}}O-SO_2-CF_3 \ + \ O \underset{}{\overset{k_p^c}{\rightleftarrows}} \ ... -O\!\!-\!\!(CH_2)_{\overline{4}}\!\!-\!\!{}^+O \ , \ {}^-O-SO_2-CF_3 \tag{113}$$

The corresponding rate constants k_{ii} and k_{tt} as well as the related activation parameters are given in Table 15.

A comparison of the rate constants of collapse of the ion pair (temporary termination) with the rate constant of propagation shows that, at least in CH_2Cl_2 solution, both constants nearly coincide. Thus, apparently, the nucleophilicities of THF and of the anions studied are of a comparable magnitude.

5.1.6 Summary of the Kinetics of Propagation Involving Covalent and Ionic Species

The general scheme of polymerization of a heterocyclic monomer involving temporary termination includes a number of different species: covalent macromolecules, macroion pairs and macroions:

$$\tag{114}$$

$$
\begin{array}{c}
P_n^+ \ + \ A^- \ + \ M \ \underset{k_d^+}{\overset{k_p^+}{\rightleftarrows}} \ P_{n+1}^+ \ + \ A^- \\
\\
P_n^+, A^- \ + \ M \ \underset{k_d^{\pm}}{\overset{k_p^{\pm}}{\rightleftarrows}} \ P_{n+1}^+, A^- \qquad ;\ K_D \\
\\
k_{tt}^{\pm} \diagdown \quad k_{ii}^{\pm} \qquad k_p^c \diagup \quad k_{tt}' \qquad ;\ K_e = k_{ii}/k_{tt} \\
\\
P_n^c \ + \ M
\end{array}
$$

There is one more reaction not included into this scheme. This is a bimolecular collapse of ions

$$P_n^+ + A^- \ \rightleftharpoons \ P_n^c \tag{115}$$

into the covalent macromolecules (identical with the external return in Winstein's terminology[231]). This reaction has not been included because one may assume that collapse should proceed via the ion pair stage (internal return).

The complete Scheme (114) has only been described for the polymerization of THF[13]. It has been shown that $k_p^+ = k_p^{\pm} (= k_p^i)$ and this equality reduces the number of elementary reactions, that have to be independently treated. Four of the rate constants, namely k_p^i, k_p^c, k_{ii}, and k_{tt} have been determined directly. The fifth one, $k_d = k_p [M]_e$, is determined under equilibrium conditions and the sixth, k'_{tt}, can be determined by applying to Scheme (114) the principle of microreversibility. All these constants, together with the dissociation constants K_D, are listed in Table 16 for the same initial concentration of monomer and in three different solvents.

First of all it is of interest to compare the reactivities of the endocyclic and exocyclic carbon atoms of the tertiary oxonium ion. These can be conducted by using the rate constants of reactions with the $CF_3SO_3^-$ anion within an ion pair taking into account the statistical factor:

$$\ldots -CH_2-CH_2-\overset{+}{O}\text{(ring)}$$

$$k'_{tt} \diagdown \quad \diagup k_{tt}$$

$$CF_3SO_3^-$$

Thus, the endo/exo ratio of reactivities ($= k_{tt}/2\,k'_{tt}$) is in the range from 15 to 20. This higher reactivity of the endocyclic position is certainly due to the ring strain.

Another comparison is possible between a unimolecular (intramolecular) and bimolecular (involving monomer) conversion of the macroester into macroion pairs: e.g.:

$$\ldots -CH_2-O-CH_2-CH_2-CH_2-CH_2-O-SO_2-CF_3 \; \underset{\phantom{k_{ii}}}{\overset{k_{ii}}{\rightleftharpoons}} \; \ldots -CH_2-\overset{+}{O}\text{(ring)} \;\; {}^-O-SO_2-CF_3$$

$$(116)$$

and

$$\ldots -CH_2-O-CH_2-CH_2-CH_2-CH_2-O-SO_2-CF_3 \; + \; O\text{(ring)}$$

$$\Big\Updownarrow k_p^c$$

$$(117)$$

$$\ldots -CH_2-\overset{+}{O}\text{(ring)} \;\; {}^-O-SO_2-CF_3$$

For instance (cf. Table 16), for $[THF]_0 = 8.0$ mole $\cdot$ l^{-1} in CH_2Cl_2 at 25 °C, $k_{ii} = 1.9 \cdot 10^{-2}\,s^{-1}$ while k_p^c is only $0.017 \cdot 10^{-2}$ mole$^{-1} \cdot$ l $\cdot$ s^{-1}. This implies that the rate of ionization of a bimolecular attack of a monomer molecule becomes equal to that of the intramolecular (unimolecular) path when the monomer concentration exceeds 100 mole $\cdot$ l^{-1}. In other words, for every 100 ionizations proceeding intramolecularly, the macroester only ionizes $[M]_t$ times (e.g. only once if $[M]_t = 1.0$ mole $\cdot$ l^{-1}), when using a monomer molecule.

The knowledge of the rate constants and their dependence on the polymerization conditions allows to study more accurately the fate of the growing macromolecule. This can be described as follows for $[THF]_0 = 8$ mole $\cdot$ l^{-1} at 25 °C and two different solvents, namely CCl_4 (first number) and CH_3NO_2; thus, when the polymer – monomer equilibrium is established, an average macroion pair adds a monomer molecule every 7 s (8 s). The lifetime of a macroion is quite short in CCl_4 (8 s) and much longer in CH_3NO_2 (330 s). Thus, within these times only 1.2 monomer molecules can be added in CCl_4 but as many as 43 in CH_3NO_2; thereafter, the macroion pair is converted into a macroester. In this non-reactive form, it exists 125 s in CCl_4 and 8 s in CH_3NO_2, i.e. the time in which it would have been possible for an ion to add 18 monomer molecules in CCl_4 but only 1.1 in CH_3NO_2. After lapse of this time, the reaction sequence of propagation, conversion into the non-reactive ester and internal ionization (kinetically equivalent to reinitiation) starts again.

Table 16. Rate constants of elementary reactions in the polymerization of THF with $CF_3SO_3^-$ anion ($[THF]_0 = 8$ mole $\cdot$ l^{-1}; 25 °C) (Ref. 13)

Solvent	Bulk dielectric constant ε	K_D (mole $\cdot$ l^{-1})	Rate constant $k_p^i \cdot 10^2$ (mole$^{-1}\cdot$l$\cdot$s^{-1})	$k_p^c \cdot 10^2$ (mole$^{-1}\cdot$l$\cdot$s^{-1})	$k_d \cdot 10^2$ (s^{-1})	$k_{ii} \cdot 10^2$ (s^{-1})	$k_{tt} \cdot 10^2$ (s^{-1})	$k'_{tt} \cdot 10^2$ (s^{-1})
CCl_4	5.9	–	4.0	0.006	15.0	0.8	12.1	0.31
CH_2Cl_2	8.2	$1.5 \cdot 10^{-5}$ [a]	3.0	0.017	15.0	1.9	3.3	0.14
CH_3NO_2	20.6	$2.0 \cdot 10^{-3}$ [a]	2.4	0.05	13.0	12.0	0.3	0.01

[a] At $[THF]_0 = 7.0$ mole $\cdot$ l^{-1}

5.2 Termination by Irreversible Collapse of the Macroion Pair

The quantitative studies of the termination by recombination of ions within a macro-ion pair and determination of the corresponding rate constants became possible by application of the end-capping method[169].

The dependence of concentration of active species in the polymerization of THF as a function of monomer conversion is shown in Fig. 14a, b for two different initiators $C_2H_5AlCl_2$-α-epichlorohydrin (a) and $(C_2H_5)_3\overset{+}{O}\,BF_4^-$ (b). These curves are calculated on the basis of the original data given in Ref. 232.

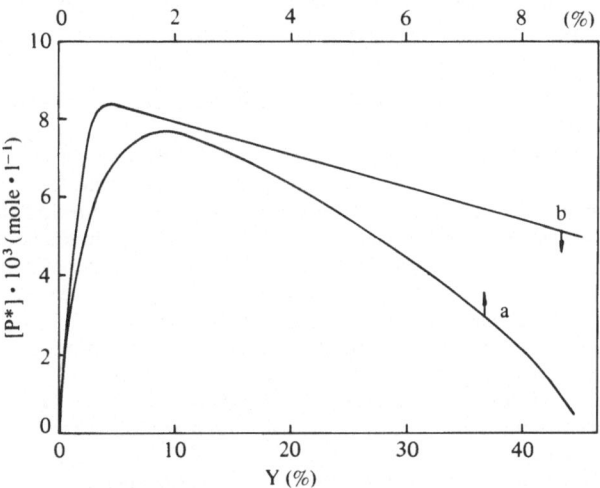

Fig. 14. Variation of concentration of active species in the polymerization of THF as a function of monomer conversion Y (%) in CH_2Cl_2 at $0\,°C$ for the systems: $[THF]_0 = 12.6$ mole $\cdot l^{-1}$, $[EtAlCl_2]_0 = [\alpha\text{-epichlorohydrin}] = 5 \cdot 10^{-2}$ mole $\cdot l^{-1}$ (a) and $[THF]_0 = 6.3$ mole $\cdot l$, $[(C_2H_5)_3\overset{+}{O}\,BF_4^-]_0 = 9 \cdot 10^{-3}$ mole $\cdot l^{-1}$ (b)

The termination reactions, responsible for the decrease of [P*], involve attack of the complex anion on the growing species followed by decomposition of the anion and formation of the haloalkyl end group:

$$\ldots-CH_2-\overset{+}{O}\diagdown\diagup\quad MtX_{n+1}^- \;\underset{(?)}{\overset{\longrightarrow}{}}\; \ldots-CH_2-O-CH_2-CH_2-CH_2-CH_2-X \;+\; MtX_n \tag{118}$$

This and similar reactions have been treated as the irreversible ones; however, the reversible process cannot be excluded at least for some MtX_n which are strong enough electrophiles. The mechanism of this opposing reaction can be presented as follows for the polymerization of THF:

$$
\begin{array}{ccc}
\underset{\substack{\text{H}_2\text{C}\diagup\overset{\text{H}_2}{\text{C}}\diagdown\text{CH}_2 \\ \text{O}\overset{+\delta}{\underset{\diagup\diagdown}{\;\cdots\;\text{C}}}\cdots\text{X}\cdots\overset{-\delta}{:}\text{MtX}_n \\ \text{H}\;\;\text{H}}}{}
& \rightleftharpoons &
\underset{\substack{\text{H}_2\text{C}\diagup\overset{\text{H}_2}{\text{C}}\diagdown\text{CH}_2 \\ \underset{\diagup\diagdown}{\overset{+}{\text{O}}\text{—C}} \\ \text{H}\;\;\text{H}}}{}\;,\; \text{MtX}_{n+1}^{-}
\end{array}
\tag{119}
$$

Mt = metal

(Anchimeric assistance of the oxygen atom from the chain)

In the case of MtX_n unable to react directly with THF (e.g. BF_3), Eq. (118) describes a real equilibrium which can be reached either after or before the monomer-polymer equilibrium is established.

Thus, the termination with complex anions may either resemble the temporary termination (Sect. 5.1.5) or it can be an irreversible process when MtX_n cannot ionize the macroalkyl halide.

Sometimes, the released metal halide MtX_n reacts further with monomer considerably complicating the overall picture.

Termination involving recombination of the growing macrocation with the anion has been responsible for a number of the interpretational errors made in the polymerization of cyclic acetals. Since, apparently, the equilibrium between macroalkyl halides (e.g. highly reactive macromolecular halomethyl ethers: $\ldots-\text{OCH}_2\text{CH}_2\text{OCH}_2\text{X}$ in the polymerization of 1,3-dioxolane) and macroions can rapidly establish and at the early stages of polymerization, the actual concentration of the growing macrocations can be equal or even less than 1% of the initiator used.

These sometimes complicated reactions involving MtX_n and monomer molecules are, in our opinion, of no interest today, after these reactions of termination and decomposition of MtX_n have been discovered and described. Some of the chemistry involving the SbCl_6^- anion studied by e.g. UV (SbCl_6^- absorbs at $\lambda_{max} = 272$ nm, $\epsilon_{max} = 1.0 \cdot 10^4$ [53, 233]), this absorption was first attributed erroneously to the oxonium macroion pair[234]) can be found in Ref. 233. We will not discuss these and further works[235, 236] involving the chemistry of SbCl_5/1,3-dioxolane interactions because, in order to obtain either high polymers or clean systems in the study of elementary reactions, the use of this anion must be avoided. Nevertheless, the reader may also consult a series of papers in which it has been shown that various compounds formed from SbCl_5 and 1,3-dioxolane or poly-1,3-dioxolane are at equilibrium and that this equilibrium can be shifted to the cationically active species by the addition of Cl^--anion donating compounds, e.g. chloromethyl methyl ether ($\text{CH}_3\text{OCH}_2\text{Cl}$):

$$
\ldots-\text{OCH}_2\text{CH}_2\text{OCH}_2\overset{+}{\text{O}}\bigcirc\;\text{SbCl}_6^- \;\rightleftharpoons\; \ldots-\text{OCH}_2\text{CH}_2\text{OCH}_2\text{Cl} + \text{SbCl}_5 + \text{O}\bigcirc
\tag{120}
$$

$$
\ldots-\text{OCH}_2\text{CH}_2\text{OCH}_2\,\text{OCH}_2\text{CH}_2\text{O}-\ldots + \text{SbCl}_5 \rightleftharpoons
$$
$$
\rightleftharpoons \ldots-\text{OCH}_2\text{CH}_2\text{OCH}_2\text{Cl} + \text{Cl}_4\text{SbOCH}_2\text{CH}_2\text{O}-\ldots
\tag{121}
$$

The number of the growing species increases due to the shifted equilibria; this was amply manifested by measuring the rates of polymerization[236].

5.2.1 Kinetics of Polymerization Involving Unimolecular Unopposed (Irreversible) First-Order Termination

In principle, the knowledge of the momentary concentrations of the active species is not necessary to determine the rate constants (cf. Fig. 14) because both k_p and k_t can be simultaneously determined by solving equations describing the non-stationary kinetics, as was shown for the cationic polymerization of styrene by Pepper[237] (partial solution) and for the polymerization of α-epichlorohydrin[238] (complete kinetic solution).

For fast initiation and unimolecular termination (first-order relative to active species) we have:

$$P_n^* + M \overset{k_p}{\underset{k_d}{\rightleftharpoons}} P_{n+1}^*; \quad -\frac{d[M]}{dt} = k_p\,([M] - [M]_e)\,[P^*]; \tag{122}$$

$$P_n^* \overset{k_t}{\longrightarrow} P_n; \quad -\frac{d[P^*]}{dt} = k_t\,[P^*]; \tag{123}$$

Thus

$$-\frac{d[P^*]}{[P^*]} = k_t \cdot dt, \text{ and } [P^*] = [P^*]_0 \exp(-k_t t); \tag{124}$$

and eventually:

$$\int_{[M]}^{[M]_0} \left(\frac{d[M]}{[M] - [M]_e}\right) = k_p\,[P^*]_0 \int_t^0 \exp(-k_t t)\,dt \tag{125}$$

After integration:

$$\ln\left(\frac{[M]_0 - [M]_e}{[M] - [M]_e}\right) = \frac{k_p}{k_t}\,[P^*]_0\,\{1 - \exp(-k_t \cdot t)\} \tag{126}$$

Therefore

$$\ln\left(\frac{[M]_0 - [M]_e}{[M]_\infty - [M]_e}\right) = \frac{k_p}{k_t}\,[P^*]_0 \tag{127}$$

$[M]_\infty$ denotes limiting monomer concentration at $t \to \infty$.

After substitution of Eq. (127) into Eq. (126) and rearrangement we have[238]:

$$\ln\left\{1 - \frac{\ln\left(\dfrac{[M]_0 - [M]_e}{[M] - [M]_e}\right)}{\ln\left(\dfrac{[M]_0 - [M]_e}{[M]_\infty - [M]_e}\right)}\right\} = -k_t \cdot t \tag{128}$$

Thus, the value of k_t can be calculated directly from Eq. (128). The knowledge of $[P^*]_0$ (for fast and quantitative initiation $[P^*]_0 = [I]_0$) enables the determination of k_p from Eq. (127). This approach has been successfully applied to the polymerization of α-epichlorohydrin[238].

5.2.2 Termination by Recombination Within an Ion Pair and Structure of Anions

The mechanism of termination by collapse of an ion pair can be considered as a nucleophilic substitution, formally similar to that of the propagation step. Thus, there are two nucleophilic attacks that have to be considered:

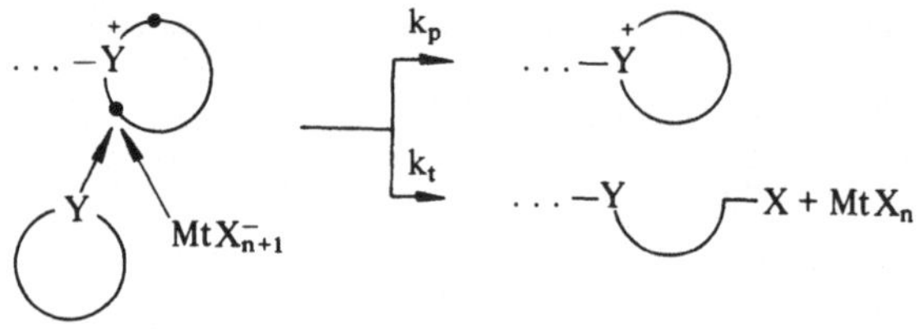

$$(129)$$

The importance of termination can be expressed by the ratio of k_t/k_p, depending on the difference of nucleophilicities of monomer and anions as well as on the inherent stability of the anion.

Thus, the rate of termination by recombination of ions increases for a given

anion with decreasing nucleophilicity of the monomer $Y\bigcirc$. This is demonstrated

below for two monomers, tetrahydrofuran ($pK_a = -2.1$) and 1,3-dioxolane ($pK_a = -4.6$) whose nucleophilicities are very different. The corresponding anions can be grouped as shown below:

Monomer	I	II	III
	AsF_6^-, SbF_6^-, PF_6^-	$SbCl_6^-$, BF_4^-	$AlCl_4^-$
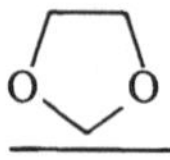	AsF_6^-, SbF_6^-	PF_6^-	$SbCl_6^-$, BF_4^-

Anions belonging to group I give no termination (or no temporary termination described by equilibria (120)–(121)) within a time necessary to reach the monomer-polymer equilibrium. In group II the termination is noticeable whereas in group III termination dominates. Thus, for instance, in the polymerization of 1,3-dioxolane with $SbCl_6^-$ or BF_4^- anions, only about 1% of initiator bearing these anion (e.g. $(C_2H_5)_3O^+$ BF_4^- or $SbCl_6^-$) is converted into the ionic active species bearing the same anions, the rest existing as various non-active covalent species equilibrated with the ions.

With more basic monomers the importance of termination decreases, as has already been discussed for the polymerization with initiators bearing simple anions (e.g. halide anion, Sect. 3). Thus, two factors have to be taken into account simultaneously, the nucleophilicity and inherent stability of the anion as well as the nucleophilicity of the monomer involved. The higher the nucleophilicity of the monomer, the less important becomes termination involving anions.

Apparently, the polymerization of some three-membered cyclic ethers, for example α-epichlorohydrin[238, 239] which undergoes incomplete monomer conversion, proceeds with efficient decomposition of the macroin pairs. It seems, that the limited conversions observed in Ref. 239 are also due to termination and not to the thermodynamic equilibrium, proposed in the cited paper and leading to the unrealistically low ring strain of α-epichlorohydrin.

5.3 Transfer and Termination Involving the Polymer Backbone

In the majority of cationic polymerizations of heterocyclic monomers, chain propagation involves a nucleophilic attack of a monomer molecule on the strained onium ions, e.g.:

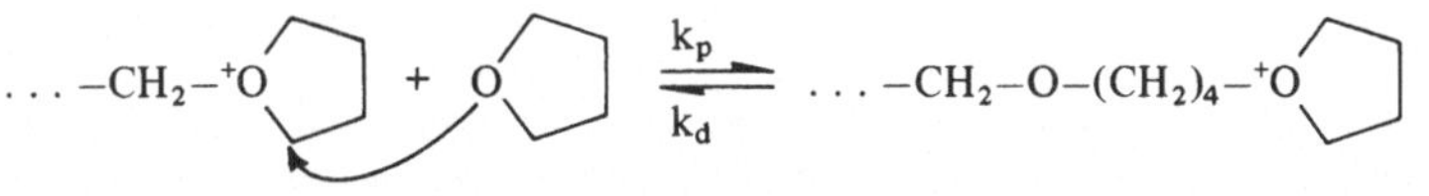

$$\dots -CH_2-{}^+O \quad + \quad O \quad \underset{k_d}{\overset{k_p}{\rightleftharpoons}} \quad \dots -CH_2-O-(CH_2)_4-{}^+O \qquad (130)$$

In the propagation step the heteroatom of a monomer molecule is engaged in the attack and, as a result of this reaction, the polymer segment including the same heteroatom is formed. Thus, it is reasonable to assume that the polymer segments will also attack the growing species. This can be a polymer segment reacting either with an active species of the foreign macromolecule (131) or with an active species of its own macromolecule (back-biting) (132):

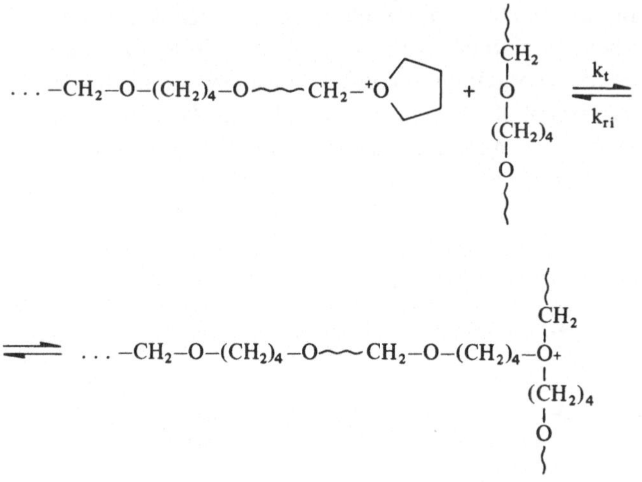

$$\dots -CH_2-O-(CH_2)_4-O\!\!\sim\!\!\sim\!\!CH_2-{}^+O \quad + \quad \underset{\substack{|\\(CH_2)_4\\|\\O}}{\overset{\substack{\{\\CH_2\\|\\O}}{}} \quad \underset{k_{ri}}{\overset{k_t}{\rightleftharpoons}}$$

$$(131)$$

$$\rightleftharpoons \quad \dots -CH_2-O-(CH_2)_4-O\!\!\sim\!\!\sim\!\!CH_2-O-(CH_2)_4-\overset{+}{O}$$

$$\ldots-CH_2-O-(CH_2)_4-O\sim CH_2-{}^+O\langle\,\rangle \underset{k_{ri}}{\overset{k_t}{\rightleftharpoons}} \ldots-CH_2-{}^+O\begin{matrix}(CH_2)_4-O\\ \diagup\qquad\quad\diagdown\\ \qquad\qquad CH_2\\ \diagdown\qquad\quad\diagup\\ (CH_2)_4-O\end{matrix} \qquad (132)$$

Both reactions, namely the intermolecular process (131) and the intramolecular one (132) can be either reversible or irreversible (termination). In the case of reversible reactions true chain transfer takes place when the rate constant of the backward reaction (k_{ri}) becomes comparable with the rate constant of propagation. This applies to the polymerization of cyclic acetals where the product of chain transfer is equally active in propagation.

Thus, transfer or termination which involve reaction of the growing center with its own backbone are inherent features of the cationic polymerization of heterocyclic monomers, and cannot be completely avoided. (For exceptions and their origin see p. 104.)

5.3.1 Evidence of Reactions with Polymer Chains

The first observations of chain transfer come from studies of cyclic oxides. As described in Sect. 5.1, the formation of a cyclic dimer (1,4-dioxane) was observed in the polymerization of ethylene oxide[240], and involves intramolecular attack within a penultimate unit of the chain:

$$\ldots-O-CH_2-CH_2-O-CH_2-CH_2-{}^+O\langle\,\rangle \rightleftharpoons \ldots-O-CH_2-CH_2-{}^+O\langle\,O\,\rangle \qquad (133)$$

This reaction is practically irreversible and the release of 1,4-dioxane from its ionized form can proceed either by an attack of the oxygen atom of its own chain on the exocyclic carbon atom of the oxonium ion, producing a new dioxanium ion at the chain end or by a similar attack of a foreign macromolecule. This process would result in depropagation to the dimer. 1,4-Dioxane is also formed when the preformed polymer is treated with an initiator of polymerization in the absence of ethylene oxide.

Other evidence for the formation of macrocycles by similar mechanisms comes from the studies of the polymerization of oxetane and 3,3-dimethyloxetane. It was demonstrated by Rose[115,241] and confirmed by Dreyfuss[242], Worsfold[243] and Goethals[244] that cyclic trimers and tetramers and higher cyclic oligomers are formed from these two monomers in polymerization.

Other evidence stems from ^{1}H-NMR studies of the polymerization of cyclic sulfides[140].

The application of 300 MHz ^{1}H-NMR to the polymerization of 3,3-dimethylthietane initiated with trialkyloxonium tetrafluoroborate (initiator concentration higher than 0.02 mole $\cdot$ l^{-1}) permits polymeric sulfonium ions to be observed:

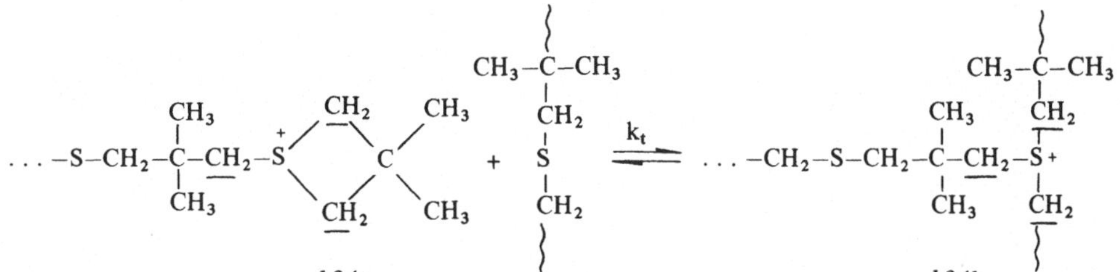

$$
\ldots -S-CH_2-\overset{\overset{\textstyle CH_3}{|}}{\underset{\underset{\textstyle CH_3}{|}}{C}}-CH_2-\overset{+}{S} \Big\langle \quad + \quad S \quad \underset{k_t}{\rightleftharpoons} \quad \ldots -CH_2-S-CH_2-\overset{\overset{\textstyle CH_3}{|}}{\underset{\underset{\textstyle CH_3}{|}}{C}}-CH_2-\overset{+}{S}
$$

$$
134a \qquad\qquad\qquad 134b
$$

$$
{}^1H-NMR: \quad \delta = 3.80, \quad 3.66 \qquad\qquad\qquad\qquad \delta = 3.60\ ppm
$$
$$
\text{and}
$$
$$
3.94\ ppm\ (AB\ system)
$$

(134)

A quantitative agreement was found between a disappearing strained monomeric sulfonium ion *134a* and an appearing non-strained polymeric sulfonium ion *134b*. In the polymerization of 1-methylazetidine a similar mechanism of degradative transfer (termination) is based on kinetic measurements. Direct NMR observations have however not been performed[203].

Further evidence of the chain transfer (termination) involving polymer chains comes from the polymerization of a given monomer A in the presence of a preformed polymer, e.g. poly-B:

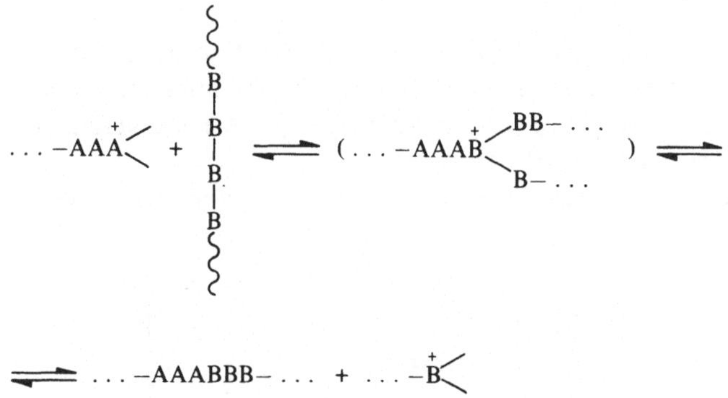

$$
\ldots -A A \overset{+}{A} \Big\langle \quad + \quad \underset{\text{(poly-B)}}{\overset{\text{B}}{\underset{\text{B}}{|}}} \quad \rightleftharpoons \quad (\ldots -AAA\overset{+}{B}\Big\langle {\overset{BB-\ldots}{\underset{B-\ldots}{}}}\ ) \quad \rightleftharpoons
$$

$$
\rightleftharpoons \quad \ldots -AAABBB-\ldots \quad + \quad \ldots -\overset{+}{B}\Big\langle
$$

(135)

where $\ldots -\overset{+}{A}\Big\langle$ and $\ldots -\overset{+}{B}\Big\langle$ are growing species.

A similar reaction of segmental exchange has been known for a long time in the chemistry of polysulfides[245]; the first observation of this reaction in the cationic polymerization of heterocycles involved polymerization of acetals[246] which then has extensively been studied[154, 247, 248].

Block copolymers are formed at the beginning of interaction[249]; the further reaction leads to the more pronounced reshuffling which eventually gives the statistically equilibrated distribution of segments.

Finally, these reactions led to a new method of preparation of copolymers based on the *interaction of two homopolymers in the presence of a cationic initiators* (schematically):

$$
\begin{array}{cccccccc}
\ldots-\mathrm{A} & & \mathrm{B}-\ldots & & \ldots-\mathrm{A} & & \mathrm{B}-\ldots & \\
\mathrm{A} & & \mathrm{B} & & \mathrm{A} & & \mathrm{B} & \\
\mathrm{A} & & \mathrm{B} & & \mathrm{A} & & \mathrm{B} & \\
\mathrm{A} & +\ \mathrm{R}^{+}\ + & \mathrm{B} & \rightleftharpoons & {}^{+}\mathrm{A}-\mathrm{R} & +\ & \mathrm{B} & \rightleftharpoons \\
\mathrm{A} & & \mathrm{B} & & \mathrm{A} & & \mathrm{B} & \\
\mathrm{A} & & \mathrm{B} & & \mathrm{A} & & \mathrm{B} & \\
\ldots-\mathrm{A} & & \mathrm{B}-\ldots & & \ldots-\mathrm{A} & & \mathrm{B}-\ldots &
\end{array}
$$

$$(136)$$

$$
\begin{array}{ccccccc}
 & & \mathrm{B}-\ldots & & \mathrm{B}-\ldots & & \\
\ldots-\mathrm{A} & & \mathrm{B} & & \mathrm{B} & & \mathrm{A}-\mathrm{R} \\
\mathrm{A} & & \mathrm{B} & & \mathrm{B} & & \mathrm{A} \\
\mathrm{A} & +\ \mathrm{A}-\mathrm{B}^{+} & & \rightleftharpoons & \mathrm{B} & +\ {}^{+}\mathrm{A}-\mathrm{A}-\ldots \\
\mathrm{A}-\mathrm{R} & \mathrm{A}\ \ \mathrm{B} & & & \mathrm{A} & & \mathrm{B} \\
 & \ldots-\mathrm{A}\ \ \mathrm{B} & & & \mathrm{A} & & \mathrm{B} \\
 & \mathrm{B}-\ldots & & & \ldots-\mathrm{A} & & \mathrm{B}-\ldots
\end{array}
$$

Scheme (136) shows only the principle of these reactions observed for a number of systems; for instance, exchange between two different polyacetals, polyethers, polyether, and polyester etc.[248,250].

One of the well documented examples for these exchange reactions is due to studies of the polymerization of propylene sulfide[251]. A mixture of non-deuterated and deuterated preformed polymers gives a copolymer, in which the deuterated and non-deuterated units are distributed at random.

This can be visualized, in agreement with the reactions discussed above, as multiple transalkylation:

$$
\begin{array}{ccccc}
\vdots & \vdots & & \vdots & \vdots \\
\mathrm{CH}_2 & \mathrm{CH}_2 & & \mathrm{CH}_2 & \mathrm{CH}_2 \\
\mathrm{CH}_3-\mathrm{CD} & \mathrm{CH}_3-\mathrm{CH}\quad \mathrm{CH}_3 & & \mathrm{CH}_3-\mathrm{CD} & \mathrm{CH}_3-\mathrm{CH} \\
\mathrm{S}\!: & {}^{+}\mathrm{S}-\mathrm{CH}_2-\mathrm{CH}-\ldots & \rightleftharpoons & {}^{+}\mathrm{S}\!-\!\!-\!\mathrm{CH}_2 & \mathrm{S} \quad(137) \\
\mathrm{CH}_2 & \mathrm{CH}_2 & & \mathrm{CH}_2\quad \mathrm{CH}-\mathrm{CH}_3 & \mathrm{CH}_2 \\
\mathrm{CH}_3-\mathrm{CD} & \mathrm{CH}-\mathrm{CH}_3 & & \mathrm{CH}_3-\mathrm{CD}\quad\mathrm{S} & \mathrm{CH}-\mathrm{CH}_3 \\
\mathrm{S} & \mathrm{S} & & \mathrm{S} & \\
\vdots & \vdots & & \vdots & \vdots
\end{array}
$$

The formation of mixed tetramers containing both deuterated and non-deuterated units was also observed when a mixture of deuterated and non-deuterated linear polymers was treated as described above.

All these examples show that polymers with heteroatoms react in various ways with cationic species, either intramolecularly or intermolecularly.

These and similar interchain exchange reactions have been discussed in more detail by Enikolopyan, Irzhak and Rozenberg in the recently published book on the interchain exchange of macromolecules [154].

5.3.2 Intermolecular Chain Transfer (Termination) to Polymer Chains

Direct ^{1}H-NMR evidence of the transformation of the strained cyclic onium ions into the polymeric onium ions in the reaction involving growing species and macromolecules was discussed in Sect. 5.3.1 for 3,3-dimethylthietane. However on the basis of ^{1}H-NMR one cannot distinguish whether the reaction proceeds intra- or intermolecularly but the answer can be found by kinetic studies.

The first complete kinetic treatment of an intermolecular interaction was given in the studies of the polymerization of 3,3-bis-(chloromethyl)oxetane [122]. The following kinetic scheme describes the non-stationary process in which termination on polymer competes with chain propagation (initiation is assumed to be fast):

$$I + M \xrightarrow[\text{fast}]{k_i} P_1^*$$

$$P_1^* + M \xrightarrow{k_p} P_2^* \tag{138}$$

$$P_n^* + M \xrightarrow{k_p} P_{n+1}^* \tag{138a}$$

$$P_n^* + P_m \xrightarrow{k_t} P_n^* P_m \text{ (inactive)} \tag{138b}$$

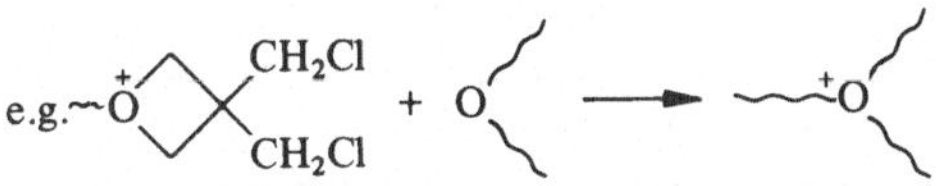

e.g.

Since

$$-d[M]/dt = k_p [M][P^*] \tag{138a'}$$

and

$$d[P^*]/dt = -k_t[P^*]([M]_0 - [M]) \tag{138b'}$$

(for fast initiation) we have, after determination of $[P^*]$ from Eq. (138b') and introduction of

$$[P^*] = [I]_0 - k_t/k_p \cdot \left\{ [M]_0 \ln \frac{[M]_0}{[M]} - ([M]_0 - [M]) \right\} \text{ into Eq. (138a')} \text{ [252]}:$$

$$-\frac{d\ln[M]}{dt} = k_p[I]_0 - k_t \left\{ [M]_0 \ln \frac{[M]_0}{[M]} - ([M]_0 - [M]) \right\} \tag{139}$$

(in the case of the reversibility of propagation, the effective monomer concentration has to be used; thus, from $[M]_0$ and $[M]$ the equilibrium monomer concentration $[M]_e$ has to be substracted) [139].

Thus, from a plot of $\ln[M]$ vs. time, $(d\ln[M]/dt)$ is graphically determined for various times and then plotted against the variable part of the right-hand side of Eq. (139) (given in { }). k_t is obtained from the slope and k_p from the intercept of the resultant straight line.

This treatment was applied to the polymerization of 3,3-bis-(chloromethyl)-oxetane[122] (irreversible propagation) and, more recently, to the polymerization of cyclic esters of phosphoric acid[139, 253] (reversible propagation):

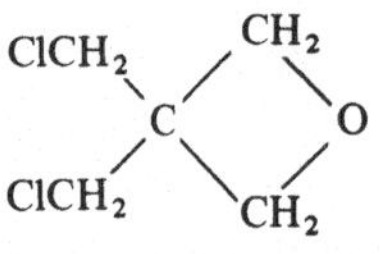

(BCMO) 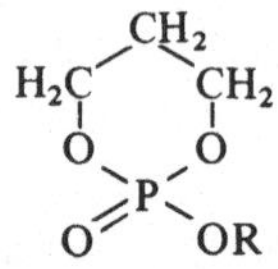 R = e.g. CH_3, C_2H_5

(2–Alkoxy–2–oxo–1, 3,
2–dioxaphosphorinanes: AODP)

In both cases appropriate straight lines were obtained permitting the determination of k_p and k_t. In the polymerization of BCMO
$k_p/k_t \cong 10^4$ (at 70 °C in chlorobenzene $k_p = 8.5\ \text{mole}^{-1}\cdot l\cdot s^{-1}$ $k_t = 1.2\cdot 10^{-3}\,\text{mole}^{-1}\cdot l\cdot s^{-1}$).

In the polymerization of 2-alkoxy-2-oxo-1,3,2-dioxaphosphorinanes it has been shown that although k_p does not depend too much on the structure of the substituent R, the rate constant of termination k_t decreases with increasing size of the alkyl group and for the large substituents (e.g. $R = (CH_3)_3Si$) k_t becomes eventually so small that it cannot be measured. The corresponding rate constants and activation parameters are given in Table 17.

Table 17. Rate constants of propagation and termination and corresponding activation parameters in the polymerization of 2-RO-2-oxo-1,3,2-dioxaphosphorinanes (Ref. 253)

Parameter	R		
	CH_3	C_2H_5	$(CH_3)_3Si$
$k_p \cdot 10^2$	1.0	0.82	0.54
$k_t \cdot 10^5$	3.5	0.26	a
$\Delta H_p^{\ddagger}$	20	21	19
$\Delta S_p^{\ddagger}$	-16	-12	-20
$\Delta H_t^{\ddagger}$	12	9	$-$
$\Delta S_t^{\ddagger}$	-48	-59	$-$

a Too small to be measured

k_p and k_t measured at 100 °C and given in $\text{mole}^{-1}\cdot l\cdot s^{-1}$; $\Delta H_p^{\ddagger}$ in $\text{kcal}\cdot\text{mole}^{-1}$ and $\Delta S_p^{\ddagger}$ in $\text{cal}\cdot\text{mole}^{-1}\cdot K^{-1}$

Analysis of the data of Table 17 indicates that a clearly larger decrease of k_t with the size of exocyclic group is mostly due to the increasingly negative entropy of activation. Indeed, the larger exo substituents have a larger number of degrees of freedom frozen in the transition state; this observation is in agreement with the observed ΔS_p value of the monomer-polymer equilibrium. In the termination reaction, segments of two macromolecules become immobilized; this making $\Delta S_t^{\ddagger}$ more negative than $\Delta S_p^{\ddagger}$.

The higher values of $\Delta H_p^{\ddagger}$ than $\Delta H_t^{\ddagger}$ reflect the higher nucleophilicity of the chain segment than that of the monomer.

The kinetic treatment leading eventually to Eq. (139) also describes the following dependence of the proportion of the originally formed active species (P_n^* in Scheme (138) still alive (for an irreversible propagation) at the monomer concentration [M]:

$$\frac{[P^*]}{[I]_0} = 1 - \frac{k_t/k_p}{[I]_0} \left\{ [M]_0 \ln \frac{[M]_0}{[M]} - ([M]_0 - [M]) \right\} \tag{140}$$

(where $[I]_0 = [P^*]_0$)

The dependence of $[P^*]/[I]_0$ on $([M]_0 - [M])/[M]_0$ is described by Fig. 15 for different $(k_t/k_p)/[I]_0$ values (shown directly on the corresponding curves[254]).

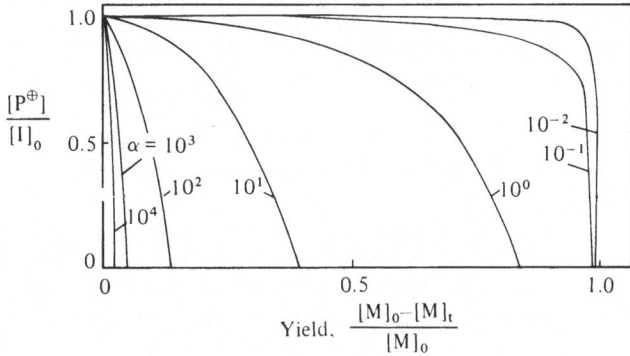

Fig. 15. Dependence of the proportion of the originally formed active centers ($[P^*]/[I]_0$) on conversion ($([M]_0 - [M])/[M]_0$) for different values of $(k_t/k_p)/[I]_0$

Usually, $[I]_0$ is of the order of 10^{-4} mole $\cdot$ l^{-1}. Thus, if $[M]_0 = 1.0$ mole $\cdot$ l^{-1} and $k_t/k_p \approx 10^{-4}$ (conditions used for BCMO polymerization)[122], then the limited conversion, when all of the active species are transformed into the $P_n^* P_m$ species (considered inactive), should be close to 0.8. As it was reported, the kinetic curves indeed level off (rates approach zero) at the plateau corresponding to approx. 80% of conversion under these conditions. Since the slope of the "plateau" is not perfectly equal to zero, there is probably some reinititation which has not been taken into account in the formulation of the kinetic Scheme (138).

Kern and Jaacks[255] and independently Enikolopyan[256] determined the ratio k_t/k_p for the polymerization of 1,3,5-trioxane as being close to 1.0. This high value is attributed to the known fact that the linear polyacetals are more basic than the cyclic ones. In these systems, the interaction of the growing species with the polymer segments is certainly a reversible process: thus, these systems cannot be quantitatively described by the kinetic Scheme (138). Nevertheless, if reversibility is tentatively neglected, then for $[M]_0 = 1.0$ mole $\cdot$ l^{-1} and $[I]_0 = 10^{-4}$ mole $\cdot$ l^{-1} the complete conversion of the originally formed active species would take place only after 1.4% of conversion of monomer into polymer.

If reaction $P_n^* + P_m \rightarrow P_n^* P_m$ (Scheme (138)) is irreversible, then limited conversion of monomer should result, provided that k_t/k_p is sufficiently large and $[I]_0$ sufficiently low. The dependence of the limit of conversion $L = ([M]_0 - [M]_\infty)/[M]_0$ on the starting monomer concentration for various values of $\alpha = (k_t/k_p)/[I]_0$ was computed and it was shown that these limited conversions decrease with increasing starting monomer concentration $[M]_0$[254].

A similar treatment as outlined for BCMO (p. 101), a four-membered cyclic ether, has later been applied by Goethals to four-membered cyclic sulfides and cyclic amines. The k_p/k_t ratios were determined for a series of substituted thietanes and azetidines; monomer structures and determined values of k_p/k_t are given in the following order:

S : 4.0 ;   X S : 20 ;   X S : 360 ;

N–CH$_3$: 250 ;   N–CH$_3$: very large

This series also shows, as was discussed for substituted 2-alkoxy-2-oxo-1,3,2-dioxaphosphorinanes that substitution influences k_t to the larger extent than k_p. Apparently, in all these systems the entropy of activation decreases with substitution more rapidly for termination than for the propagation reaction.

5.3.3 Chain Transfer to Polymer in the Polymerization of THF

Chain transfer to poly-THF in the polymerization of THF has never been systematically studied. Nevertheless, there are a few reports in which the rate constant of transfer to polymer was estimated or at least provided data making this estimation possible.

Chain transfer to polymer in the polymerization of THF is discussed separately, because THF is the model monomer for which the polymerization is best understood. This transfer leads to the transformation of cyclic, strained tertiary oxonium ions into non-strained oxonium ions, linear or macrocyclic, e.g. (for the intermolecular transfer resulting in a linear structure):

$$\ldots -CH_2-{}^+O \quad + \quad O \begin{smallmatrix} CH_2-CH_2- \ldots \\ CH_2-CH_2- \ldots \end{smallmatrix} \quad \underset{k_{ri}}{\overset{k_{tr}}{\rightleftharpoons}}$$

$$\rightleftharpoons \quad \ldots -CH_2-O-(CH_2)_4-\overset{+}{O} \begin{smallmatrix} CH_2-CH_2- \ldots \\ CH_2-CH_2- \ldots \end{smallmatrix}$$

(141)

This reaction is manifested by the following changes in the system:

1) The α and β methylene CH$_2$ groups (endo- and exocyclic ones) change their chemical environment. In the new environment their chemical shifts in the ^{1}H- and ^{13}C-NMR spectra differ from the previous ones. The corresponding spectra cannot, however, discriminate between the intermolecular and intramolecular (back-biting) chain transfer, because in the non-strained rings, formed by back-biting,

the chemical shifts of the corresponding atoms are the same as in the non-strained non-cyclic structures.

2) The viscosity of the living system containing oxonium ions at the junction points of two macromolecules, should be higher than for the same system, after decomposition, which annihilates the junction points.

3) Chain transfer to polymer is usually followed by reinitiation (k_{ri} in Eq. (141)) and the corresponding rupture of the macromolecules leads to the redistribution of the polymerization degrees, changing in a characteristic manner the $\overline{M}_w/\overline{M}_n$ ratio (broadening). The rate of this broadening gives an access to k_{tr}.

These three phenomena, associated with chain transfer to polymer, were indeed observed in the cationic polymerization of THF.

Pruckmayr observed that in the polymerization of THF in CH_3NO_2 the decrease of the concentration of strained cyclic tertiary oxonium ions is counterbalanced by the formation of the non-strained oxonium ions[257]. The change was observed at 25 °C by ^{1}H-NMR and 6 h after monomer-polymer equilibrium was established some 7% of the original oxonium ions became converted.

The increase of the viscosity of living poly-THF, brought to the living polymer-monomer equilibrium, was reported by Rosenberg and Lyudvig[258]. This increased viscosity can be drastically reduced by the addition of small amounts of KOH which is known to annihilate oxonium ions. Gelification of the system, probably due to the formation of the infinite network containing oxonium junction points, was noted by Franta and Rempp in the cationic grafting of THF onto poly(vinyl chloride)[259].

There are two reports describing the changes of $\overline{M}_w/\overline{M}_n$ in the polymerization of THF. The observed broadening is larger than could have been estimated if only the depropagation is taken into account. This broadening has been attributed to chain transfer to polymer. Enikolopyan a.o. calculated $k_{tr}/k_p \simeq 10^{-4}$ at 20 °C (thus, $k_{tr} \simeq 2 \cdot 10^{-6}$ mole$^{-1} \cdot$ l $\cdot$ s^{-1})[260] whereas Entelis found a larger value: $k_{tr} = 8 \cdot 10^{-5}$ mole$^{-1} \cdot$ l $\cdot$ s^{-1} [261]. In the latter study, polymerization was initiated by the BF_3 propylene oxide system which is known to give some termination. This can be responsible for additional broadening and thus higher k_{tr} values.

The above estimated values can be compared with the results of the model system, namely the exchange reaction between alkylated THF and diethyl ether studied by Saegusa[87]:

$$C_2H_5-{}^+O\!\!\!\overset{\displaystyle\frown}{\underset{\displaystyle\smile}{}} + O\!\!\!<\!\!\!{\overset{\displaystyle CH_2-CH_3}{\underset{\displaystyle CH_2-CH_3}{}}} \underset{k'_{ri}}{\overset{k'_{tr}}{\rightleftharpoons}} C_2H_5-{}^+O\!\!\!<\!\!\!{\overset{\displaystyle \underset{\displaystyle CH_2}{\overset{\displaystyle CH_3}{|}}\,|}{\underset{\displaystyle \underset{\displaystyle CH_3}{\overset{\displaystyle CH_2}{|}}\,|}{}}} + O\!\!\!\overset{\displaystyle\frown}{\underset{\displaystyle\smile}{}} \qquad (142)$$

Actually, Seagusa[87] studied the reversible alkylation of THF by triethyloxonium salt and obtained in CH_2Cl_2 at 35 °C $k'_{tr} = 4.2 \cdot 10^{-5}$ mole$^{-1} \cdot$ l $\cdot$ s^{-1}. This value is not very far from that estimated by Enikolopyan a.o.

The equilibrium constant in Eq. (142) determined by Saegusa is equal to $K = k'_{tr}/k'_{ri} = 2.8 \cdot 10^{-2}$.

Thus, we can conclude that k_{tr} in the polymerization of THF, although not yet known with sufficient certainty, is probably close to 10^{-5} mole$^{-1} \cdot$ l $\cdot$ s^{-1}. Under typical polymerization conditions, e.g. in CH_2Cl_2 at 25 °C and $[THF]_0 = 8.0$ mole $\cdot$ l^{-1} and $[P^*] = 10^{-2}$ mole $\cdot$ l^{-1}, less than 3% of the active species (strained cyclic oxonium ions) will be converted into the non-propagating oxonium ions when polymerization attains the living polymer-monomer equilibrium. For higher concentrations of active species (e.g. 10^{-1} mole $\cdot$ l^{-1}) within a time necessary to achieve the living polymer-monomer equilibrium, this value will be smaller.

5.3.4 Intermolecular Chain Transfer Followed by Reinitiation

Reactions discussed in Sect. 5.3.1 lead to the formation of branched polymeric ions (*143*):

$$
\ldots-A-A-A-A-A\overset{+}{<} \;+\;
\begin{array}{c} A-\ldots \\ | \\ A \\ | \\ A \\ | \\ \overset{+}{A}-\ldots \end{array}
\;\rightleftharpoons\;
\ldots-A-A-A-A-A-\overset{+}{}
\begin{array}{c} A-\ldots \\ | \\ A \\ | \\ A \\ | \\ A-\ldots \end{array}
\tag{143}
$$

(the counterion is omitted; $-\overset{+}{A}<$ denotes a cyclic or branched onium ion)

These ions were considered as non-reactive, i.e. it was assumed that any reinitiation reaction is slow as compared with chain propagation and termination reactions.

There are, however, systems in which these branched polymeric ions are only dormant species, i.e. being unreactive by themselves, they are able to convert into reactive species either by reaction with monomer or with a polymeric segment. Some examples are given in the following:

1. Reinitiation by reaction with monomer (denoted by a in this scheme):

$$
\ldots-A-A-A-A-A-\overset{+}{A}
\begin{array}{c} A-\ldots \\ | \\ | \\ A \\ | \\ A-\ldots \end{array}
\;+\; a \;\xrightarrow{\;k_{rim}\;}\;
\ldots-A-A-A-A-A-\overset{+}{A}< \;+\;
\begin{array}{c} A-\ldots \\ | \\ A \\ | \\ A-\ldots \end{array}
\tag{144}
$$

k_{rim} = rate constant of reinitiation with monomer

2. Reinitation with a polymer segment, e.g.:

$$
\ldots-A-A-A-A-A-\overset{+}{A}
\begin{array}{c} A-\ldots \\ | \\ A \\ | \\ A-\ldots \end{array}
\;\xrightarrow{\;k_{rip}\;}\;
\ldots-A-A-A-A-A-\overset{+}{A}< \;+\;
\begin{array}{c} A-\ldots \\ | \\ A \\ | \\ A-\ldots \end{array}
\tag{145}
$$

k_{rip} = rate constant of reinitiation with polymer

The first case was considered for the polymerization of BCMO[122]; it involves an equilibrium between the non-reactive polymeric cations and the strained ones:

$$
\ldots-O-CH_2-\underset{\underset{CH_2Cl}{|}}{\overset{\overset{CH_2Cl}{|}}{C}}-CH_2-\overset{+}{O}
\;
\begin{array}{c} CH_2Cl \\ \diamondsuit \\ CH_2Cl \end{array}
\;+\;
\begin{array}{c} \vdots \\ CH_2 \\ | \\ O \\ | \\ CH_2 \\ | \\ ClCH_2-C-CH_2Cl \\ \vdots \end{array}
\;\underset{k_{ri}}{\overset{k_t}{\rightleftharpoons}}
\tag{146}
$$

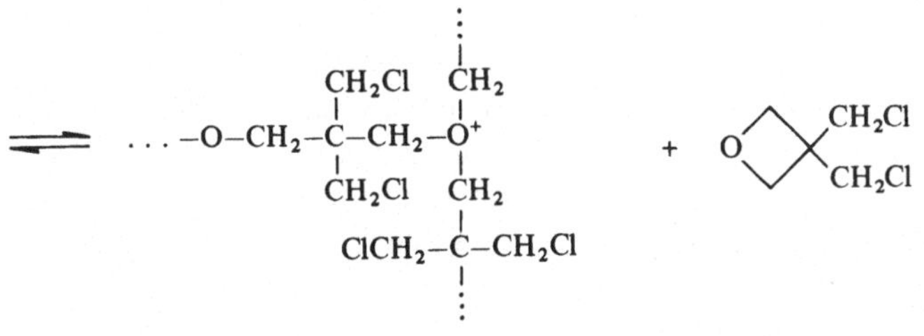

$$\rightleftharpoons \quad \ldots -O-CH_2-\underset{\underset{CH_2Cl}{|}}{\overset{\overset{CH_2Cl}{|}}{C}}-CH_2-O^+\underset{\underset{ClCH_2-\underset{|}{\overset{|}{C}}-CH_2Cl}{|}}{\overset{\overset{CH_2}{|}}{CH_2}} \quad + \quad O\diamondsuit\overset{CH_2Cl}{\underset{CH_2Cl}{\big\langle}}$$

The kinetic scheme simultaneously describing propagation, termination and reinitiation with monomer leads to kinetic equations which can be used for the determination of all three constants involved, provided that the concentration of the active species can separately be determined (e.g. by NMR).

The corresponding kinetic scheme and its solution are given below (for fast initiation):

$$P_n^* + M \overset{k_p}{\to} P_{n+1}^*$$

$$P_n^* + P_m \overset{k_t}{\to} P_n^* P_m \text{ (inactive)} \tag{147}$$

$$P_n^* P_m + M \overset{k_i}{\to} P_{n+1}^* + P_m$$

$$[P^*] + [P^*P] = [I]_0$$

$$-\frac{d[M]}{dt} = k_p[P^*][M] + k_i([I]_0 - [P^*])[M] \tag{148}$$

$$-\frac{d[P^*]}{dt} = k_t[P^*]([M]_0 - [M]) - k_i[M]([I]_0 - [P^*]) \tag{149}$$

Solution of Eq. (148) gives:

$$d \ln [M]/dt = k_i[I]_0 + (k_p - k_i)[P^*] \tag{150}$$

A graph of $d \ln [M]/dt$ against the concentration of active species according to Eq. (150) yields a straight line with $k_i[I]_0$ and $k_p - k_i$ being obtained from the intercept and slope.

If $d[P^*]/dt = 0$ (stationary state for active species), then from the slope of the plot of $([I]_0/P^* - 1)$ versus $([M]_0/[M] - 1)$ (Eq. (149)) the ratio k_t/k_i can be found. In this way, all rate constants involved, namely k_p, k_i and k_t can be determined.

Apparently, the polymerization of cyclic acetals, or at least of 1,3-dioxolane, constitutes another and highly special case where the originally formed active species (e.g. cationated monomer and/or cationated cyclic oligomers) are converted at the early stage of polymerization into the polymeric cations which are still active, e.g.:

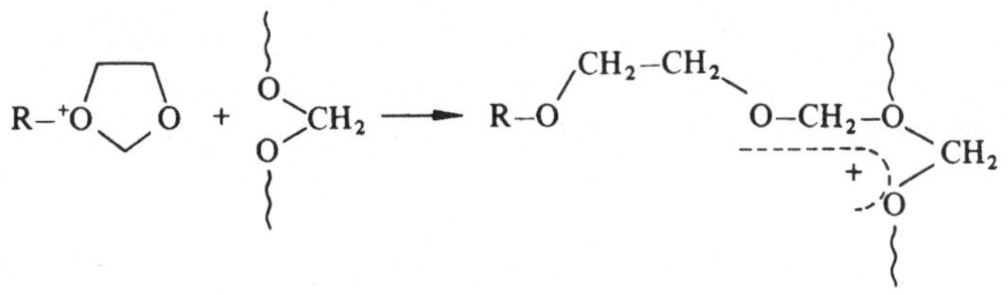

$$R-\overset{+}{O}\diagdown\diagup O \; + \; \underset{O}{\overset{O}{\diagdown}}CH_2 \; \longrightarrow \; R-O\underset{O-CH_2-O}{\overset{CH_2-CH_2}{\diagup}}\underset{-\overset{+}{-}\diagup}{\overset{\diagdown}{}}CH_2 \tag{151}$$

The propagation step involving these species (cf. however Sect. 4.1) thus requires reaction between monomer and polymeric cations, considered in all the other systems as having a very low reactivity. The main difference between the polymeric cations in polyacetals and in the polymerization of other heterocycles is due to the stabilization of the positive charge of the CH_2 group, involved in propagation, by an adjacent oxygen atom. This makes the polymeric cation like (151) a reactive growing species.

5.3.5 Intramolecular Termination

Polymerization of ethylene oxide is a suitable example of the intramolecular termination. Polymerization of other three-membered monomers like ethylene sulfides and N-substituted aziridines also terminates intramolecularly. This conclusion is based on the studies of the final monomer concentration and its dependence on the starting initiator concentration.

The corresponding kinetic scheme, including a unimolecular (intramolecular) termination, reads:

$$P_n^* + M \xrightarrow{k_p} P_{n+1}^*,$$

$$P_n^* \xrightarrow{k_t} P_n; \text{ thus } [P^*] = [I]_0 \exp(-k_t t) \tag{152}$$

thus:

$$-\int_0^t \frac{d[M]}{[M]} = k_p [I]_0 \int_0^t \exp(-k_t \cdot t) \, dt$$

The following dependence of $[M]_\infty$ on $[I]_0$ is obtained after integrating the above equations and assuming fast initiation:

$$\ln \frac{[M]_0}{[M]_\infty} = \frac{k_p}{k_t} [I]_0 \tag{153}$$

The linear plot of $\ln [M]_0/[M]_\infty$ vs. $[I]_0$ which passes through the origin indicates intramolecular termination (cf. Eq. (140), p. 103, for intermolecular termination).

The data collected by Goethals on the k_p/k_t ratio in the polymerization of cyclic amines (determined from Eq. (153)) permits to show (Fig. 16) that there is a linear relationship between $\ln (k_p/k_t)$ and the effective volume of substituents. These volumes were calculated according to Ref. 262. Thus, although the linearity itself has little importance, this dependence is in good agreement with the discussed earlier influence of substituents on the rate constants of termination and propagation.

Further substitution at the ring almost completely eliminates chain termination and provides systems resembling the living ones as has been observed in the polymerization of 2-methyl-1-(phenylmethyl)aziridine and 2-methyl-1-(2-cyanoethyl)aziridine[95].

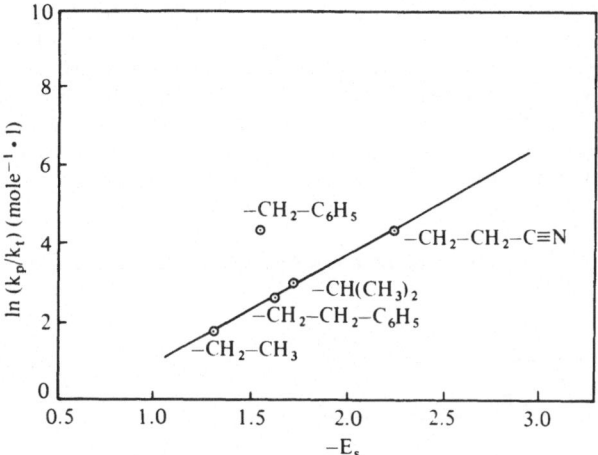

Fig. 16. Dependence of ln (k_p/k_t) on the effective volume $(-E_s)$ for N-substituted aziridines

5.3.6 Intramolecular Chain Transfer (Termination) Followed by Reinitiation

Cationic polymerization of propylene sulfide provides a good example of a process in which, at the first, non-stationary stage, the originally formed active species (strained tertiary sulphonium ions holding a monomer molecule) are converted into the dormant species (non-strained sulfonium ions) which are at equilibrium with the strained ones[41]:

(154)

(formation of tetrameric species: the 12-membered ring)

It is striking at first glance that reinitiation proceeds by folding at the chain end a highly strained three-membered ring and that reinitiation does not proceed by monomer attack. This postulate remains surprising even if it is remembered that linear sulfides are more nucleophilic than three-membered cyclic sulfides. The only explanation may be due to the assumption that the gain in entropy is sufficiently large to overcome the highly positive value that would be expected for the enthalpy of activation, being certainly higher than that for the bimolecular monomer attack.

Thus, for reinitiation proceeding as an intramolecular reaction we have the following scheme:

$$I + M \xrightarrow{fast} P_1^*$$

$$P_1^* + M \xrightarrow{k_p} P_2^* \quad (\text{irreversible})$$

$$P_n^* + M \xrightarrow{k_p} P_{n+1}^*$$

$$P_n^* \xrightarrow{k_{tt}} P_n$$

$$P_n \xrightarrow{k_{ri}} P_{n-4}^* + \text{cyclic tetramer}$$

(155)

This scheme can be solved in a straightforward way and without any assumption, leading to the following expression for $[P^*]$:

$$[P^*] = \frac{[I]_0 \{k_{ri} + k_{tt} [\exp - (k_{tt} + k_{ri})\, t]\}}{k_{tt} + k_{ri}} \tag{156}$$

Thus, when $t \to \infty$, $k_{ri} > k_{tt} [\exp - (k_{tt} + k_{ri})\, t]$ and

$$[P^*] = \frac{k_{ri}[I]_0}{k_{tt} + k_{ri}} \; .$$

Only a small portion of the originally formed active species persists in this form when equilibrium is reached. Thus, it can be assumed that $k_{tt} \gg k_{ri}$. Eventually we have

$$-d[M]/dt = k_p \cdot \frac{k_{ri}}{k_{tt}} [I]_0 \cdot [M] \, .$$

From this dependence $k_p \cdot k_{ri}/k_{tt}$ can be found.

On the other hand, at the early stages of polymerization, when reinitiation can be neglected, we have (assuming fast and quantitative initiation) the following scheme of reactions:

$$I + M \xrightarrow{k_i} P_1^* \quad (\text{fast})$$

$$P_n^* + M \xrightarrow{k_p} P_{n+1}^*$$

$$P_n^* \xrightarrow{k_{tt}} P_n$$

(157)

The solution of this scheme in a way analogous to the solution of Eq. (152) leads to Eq. (158) which can be applied to the early non-stationary stage and which gives access to both k_p and k_{tt}.

$$\frac{d \ln[M]}{dt} = k_p [I]_0 - k_{tt} \ln \frac{[M]_0}{[M]} \tag{158}$$

Once these constants are known, k_{ri} can be obtained from the values of $k_p \cdot k_{ri}/k_t$ determined as shown above.

5.3.7 Kinetics of Chain Transfer to Polymer. Determination of Rate Constants. Summary

In Table 18 various cases of chain transfer to polymer are described and the corresponding equations derived. These equations allow the involved rate constants to be determined.

5.3.8 Thermodynamics of the Linear Polymer-Macrocycle Equilibrium

The presence of macrocycles of various size in heterochain polymers prepared by polycondensation or polymerization is a well known phenomenon[263, 264]. This is due to reactions discussed in the preceding sections of this review and can be simply presented as a result of back-biting (including end-biting) reactions (the influence of end-biting will be discussed separately):

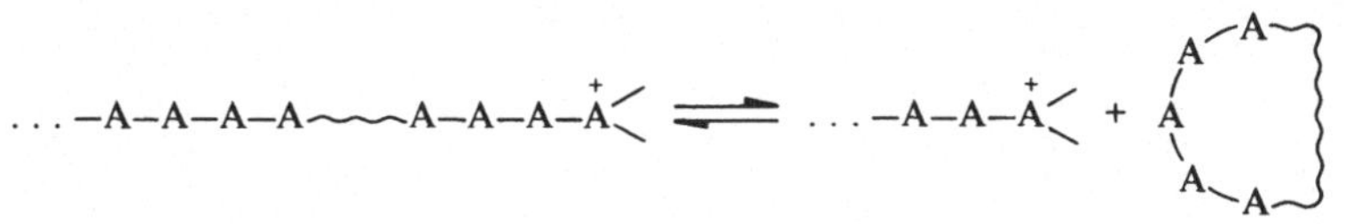

$$\ldots -A-A-A-A\sim\sim A-A-A-\overset{+}{A}\!< \;\; \rightleftharpoons \;\; \ldots -A-A-\overset{+}{A}\!< \; + \; \text{(macrocycle)} \tag{159}$$

where $-\overset{+}{A}\!<$ denotes an active center

Thus, for instance, in the polymerization of THF these processes can be visualized as follows:

Table 18. Basic kinetic equations in non-stationary kinetics with chain transfer

No	Scheme	Differential Equations	Comments
1	$I + M \xrightarrow{k_i} P*$ $M + P_n^* \xrightarrow{k_p} P_{n+1}^*$ $P* + P \xrightarrow{k_{tr}} P*P$	$-\dfrac{d[I]}{dt} = k_i[I][M]$ $-\dfrac{d[M]}{dt} = k_i[I][M] + k_p[P*][M]$ $\dfrac{d[P*]}{dt} = k_i[I]([M] - k_{tr}[P*]([M]_0 - [M])$ Solution: $-\dfrac{d\ln[M]}{dt} = k_p[I]_0 \left\{1 - \exp\left(-k_i \int_0^t [M]dt\right)\right\} - k_{tr}\left\{[M]_0\ln\dfrac{[M]_0}{[M]} - ([M]_0 - [M])\right\}$	slow initiation transfer to polymer k_i; k_p and k_{tr} can separately be determined
2	$M + P_n^* \xrightarrow{k_p} P_{n+1}^*$ $P* + P \xrightarrow{k_{tr}} P*P$	$-\dfrac{d[M]}{dt} = k_p[P*][M]$ $-\dfrac{d[P*]}{dt} = k_{tr}[P*]([M]_0 - [M])$ Solution: $-\dfrac{d\ln[M]}{dt} = k_p[I]_0 - k_{tr}\left\{[M]_0\ln\dfrac{[M]_0}{[M]} - ([M]_0 - [M])\right\}$	instantaneous initiation (schemes 2–6) transfer to polymer direct determination of k_p and k_{tr}
3	$M + P_n^* \xrightarrow{k_p} P_{n+1}^*$ $P* + P \xrightarrow{k_{tr}} P*P$ $P*P \xrightarrow{k_{ti}} P* + P$	$-\dfrac{d[M]}{dt} = k_p[P*][M]$ $\dfrac{d[P*]}{dt} = k_{ri}([I]_0 - [P*]) - k_{tr}[P*]([M]_0 - [M])$ Solution: $-\dfrac{d\ln[M]}{[I]_0 dt} + \dfrac{k_{tr}}{[I]_0}\left\{[M]_0\ln\dfrac{[M]_0}{[M]} - ([M]_0 - [M])\right\} = k_p - k_p k_{rit}$ assuming $k_{tr}[M]_0 > k_{-ri}$	transfer to polymer intramolecular reinitiation analysis of at least two experiments at constant t gives k_{tr}. For known k_{tr} both k_p and k_{ri} can be determined.

4   $M + P_n^* \xrightarrow{k_p} P_{n+1}^*$

$P^* + P \xrightarrow{k_{tr}} p^*P$

$p^*P + M \xrightarrow{k_{ti}} P^* + P$

$-\frac{d[M]}{dt} = k_p[P^*][M] + k_{ri}([I]_0 - [P^*])[M]$

$-\frac{d[P^*]}{dt} = k_{tr}[P^*]([M]_0 - [M]) - k_{ri}[M]([I]_0 - [P^*])$

Solution: $-\frac{d\ln[M]}{dt} = k_{ri}[I]_0 - (k_p - k_{ri})[P^*]$

when $d[P^*]/dt \simeq 0$; $\quad \frac{[I]_0}{[P^*]} - 1 = \frac{k_{tr}}{k_{ri}}\left(\frac{[M]_0}{[M]} - 1\right)$

transfer to polymer
reinitiation with monomer

provided that the concentration of active species can directly be determined k_p, k_{ri}, k_{tr} can be found

5   $M + P_n^* \xrightarrow{k_p} P_{n+1}^*$

$P^* \xrightarrow{k_{tr}} p^*P$

$-\frac{d[M]}{dt} = k_p[P^*][M]$

$-\frac{d[P^*]}{dt} = k_{tr}[P^*]$

Solution: $\ln\left\{1 - \ln\frac{[M]_0}{[M]} \Big/ \ln\frac{[M]_0}{[M]_\infty}\right\} = -k_{tr}\cdot t$

$\frac{1}{DP_n} = \frac{k_{tr}}{k_p} + \frac{[I]_0}{[M]_0 - [M]_\infty}$

intramolecular transfer to polymer

k_{tr} can be calculated even without knowledge of $[I]_0$ and k_p from the ratio k_{tr}/k_p

6   $M + P_n^* \xrightarrow{k_p} P_{n+1}^*$

$P^* + A \xrightarrow{k_{tr}} P^*A$

$-\frac{d[M]}{dt} = k_p[P^*][M]$

$-\frac{d[P^*]}{dt} = k_{tr}[P^*][A]$

Solution: $-\frac{d\ln[M]}{dt} = k_p[I]_0 - k_{tr}[A]_0\ln\frac{[M]_0}{[M]}$

transfer to agent A

direct determination of k_p and k_{tr}

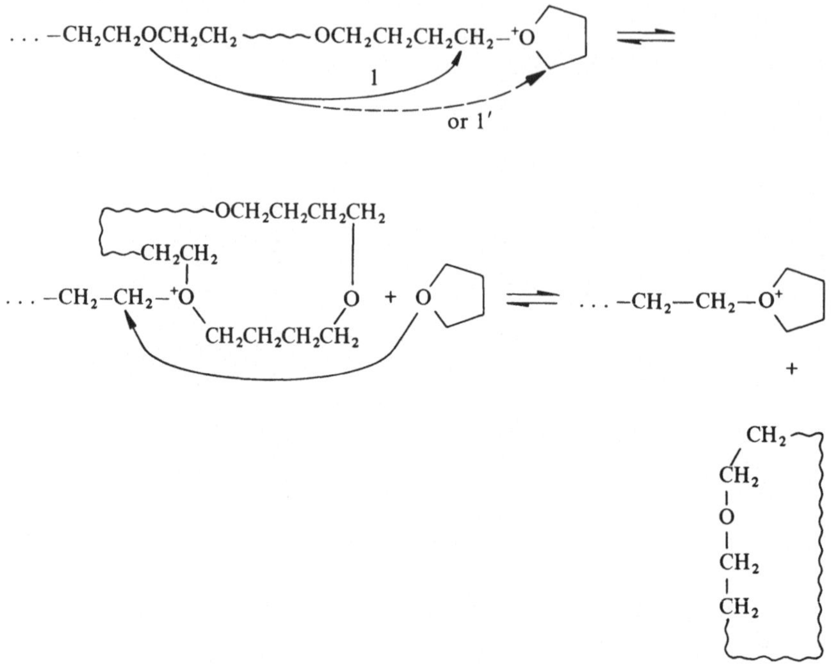

$$\ldots -CH_2CH_2OCH_2CH_2 \sim\sim\sim OCH_2CH_2CH_2CH_2-{}^+O \rightleftharpoons$$

1
or 1'

$$\ldots -CH_2-CH_2-{}^+O \ldots + O \rightleftharpoons \ldots -CH_2-CH_2-O^+ \quad (160)$$

+

$$\begin{array}{c} CH_2 \\ CH_2 \\ O \\ CH_2 \\ CH_2 \end{array}$$

The detailed mechanisms of cyclization in the polymerization of other monomers may be similar to this one or slightly different. For all these systems, however, the final equilibrium can be described by a formal scheme in which a linear polymer segment is converted into a macrocyclic segment together with a correspondingly shorter, linear segment:

$$\ldots -M_{m+n}-\ldots \longrightarrow \ldots -M_m-\ldots + \bigcirc M_n \quad (161)$$

The Jacobson-Stockmayer cyclization theory[265], which eventually relates equilibrium concentrations of macrocycles to the polymerization degree n, is based on the assumption that these concentrations are proportional to the probabilities of intramolecular cyclization. These, in turn, are given by the radii of gyration for rings corresponding to long random coil chains. This theory requires that the chains are of sufficient length and flexibility to obey Gaussian statistics.

The equilibrium constant K_n for a macrocycle of the polymerization degree n is given by

$$\left[\bigcirc M_n \right]^{-1} = K_n = \gamma \cdot n^{-5/2} \quad (162)$$

where γ is a coefficient characteristic of a given system (monomer, solvent).

According to Eq. (162) the concentration of macrocycles does not depend on the starting monomer concentration and, therefore, the proportion of cyclic macromolecules in the mixture with linear macromolecules will increase with decreasing initial monomer concentration ($[M]_0$). Thus, there should be a monomer concentration $[M]_0$ (or $[M]_0 - [M]_e$), which is equal to the sum of the equilibrium concentrations of macrocycles. Under these conditions, the macrocycles will be the only polymeric products. Thus, in principle, in any ring-opening polymerization, conditions can be found where the product of polymerization is exclusively cyclic.

5.3.9 Kinetic Enhancement in Macrocycles

Kinetic enhancement and kinetic depression has been discussed recently by the authors of this review[156, 266].

Let us imagine a macromolecule growing by a cationic mechanism and having a chain end X, formed upon initiation, and which is much more reactive (rate constant k_e) toward a cationic growing end than the heteroatoms along the chain (rate constant k_b), e.g. in the case of a polyether:

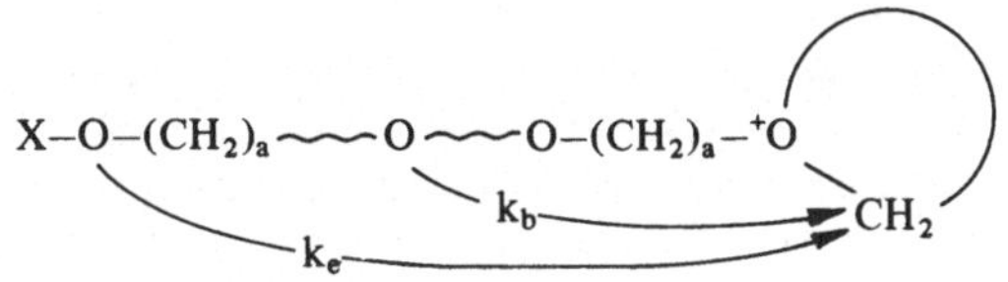

$$\tag{163}$$

Then, at equilibrium

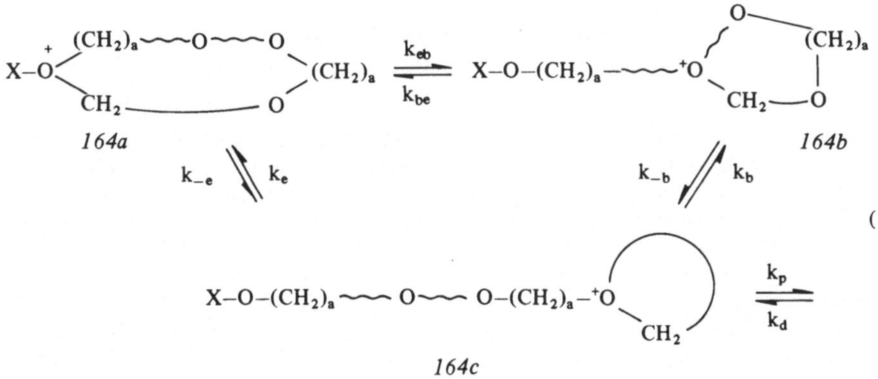

$$\tag{164}$$

where k_{eb}, k_{be}, k_e, k_{-e}, k_b and k_{-b} are rate constants of isomerization of active species.

If we now assume that $k_e \gg k_b$, and simultaneously $k_e > k_p$, and that only the strained oxonium ions can propagate (*164c*), then, at least at the early stages of polymerization, the large majority of macromolecules (the proportion given by the ratio of the corresponding rate constants) will exist in the form of the non-reactive macrocycles. If $X = H$, then the proton transfer to monomer

$$(165)$$

will dominate over other reactions of the protonated macrocycle *164a* and the final result of the equilibria described above will be the presence of the large proportion of macrocycles in the system. It is easy to imagine that the neutralization of the cations should lead to a mixture of macrocyclic and linear macromolecules with a very large excess of the cyclic ones. Thus, the concentration of macrocycles at the early stages of polymerization (small polymerization degrees) may be higher than the equilibrium concentration, reached only at the later stages of polymerization. This kinetic enhancement in macrocycles comes from the enhanced reactivity of the chain end, formed upon initiation, toward the growing cation.

Indeed, for longer chains (and, thus, for higher degrees of conversion for living systems with $k_i > k_p$) the rate constant k_{en} (which decreases with n) may become lower than the rate constants of back-biting to closer located oxygen atoms. At this point, the kinetic enhancement vanishes and the kinetically controlled proportion of macrocycles decreases, eventually reaching its thermodynamically controlled proportions.

Apparently, the polymerization of cyclic acetals, notably the polymerization of 1,3,6,9-tetraoxacycloundecane studied by Schulz[267], proceeds according to Scheme (165) with kinetic enhancement. This may be the reason that, at the early stages of polymerization, predominantly cyclic dimer, trimer, tetramer, etc., are observed although propagation proceeds on the linear growing species.

It has only recently been experimentally shown that the cyclic living (with $H-\overset{+}{O}<$ ions) and linear living macromolecules (containing hydroxymethyl end groups and $\ldots-CH_2-\overset{+}{O}<$ ions) coexist in solution of 1,3-dioxolane the polymerization of which is initiated with protonic acid and that the ratio of linear to cyclic living molecules depends on the degree of polymerization[34].

If, however, X in Eq. (164) is not a proton but e.g. an alkyl group, then its transfer becomes more difficult, and the killing reaction will provide a larger proportion of linear macromolecules. This has recently been clearly shown by Pruckmayr for the polymerization of THF with $CH_3OSO_2CF_3$ and $HOSO_2CF_3$ used as initiators[257, 269]. Although in both cases the total concentration of macrocycles is less than a few percent (it was even claimed earlier that macrocycles are absent in poly-THF[263]), in the polymerization initiated with methyl triflate, both linear and cyclic oligomers (from dimer to octamer) were observed whereas with triflic acid as initiator, no linear oligomers were isolated and only cyclic oligomers were determined (up to the octamer). This is shown in Fig. 17.

Since the oxygen atom in the primary hydroxy group is more nucleophilic than the oxygen atoms in the corresponding ethers[271], the attack of HO~~ on the growing cation (end-biting, k_e in Eq. (163)) apparently proceeds much more rapidly than similar back-biting reaction (k_b), involving ether oxygen atoms of poly-THF segments. Moreover, the macrocyclic secondary oxonium ion (*164c*) will rapidly transfer its proton to the more nucleophilic THF, while the tertiary oxonium ion (*164c*) has three equivalent sites of reaction (provided that it is not strained). The reaction with either THF molecule or some intentionally added terminating agent at two of these sites will result in linear macromolecules.

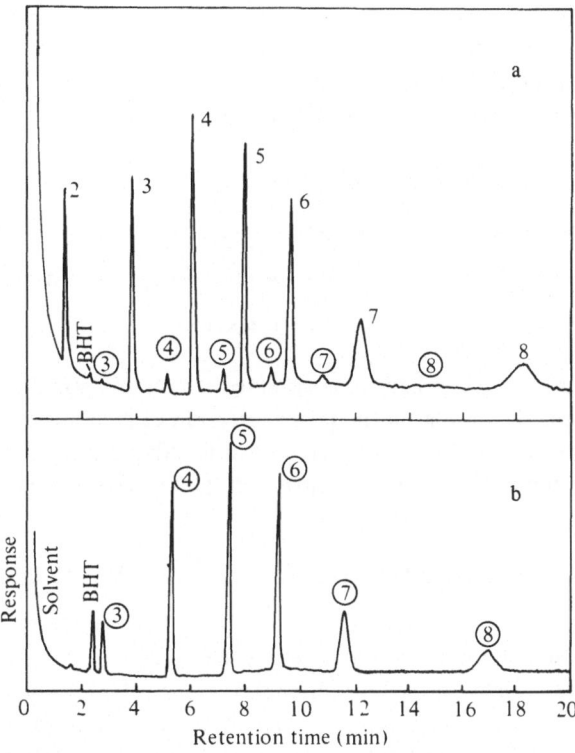

Fig. 17a, b. Gas chromatograms of a THF polymerization mixture in CH_3NO_2 initiated by $CF_3SO_3CH_3$ (a) and CF_3SO_3H (b). 2–8: linear oligomers; ③ – ⑧ cyclic oligomers (Ref. 257)

The kinetic enhancement in macrocycles is observed in the cationic polymerization of hexamethylcyclotrisiloxane $[(CH_3)_2 SiO]_3$. Since macrocycles in this system are much less reactive than monomer, their concentration increases parallelly to the concentration of high polymer up to the moment when almost all the monomer is consumed. Only oligomers with number of siloxane units being the multiple of the number in monomer (n = 6, 9, 12 . . .) appear in a large excess what permits to reject a depropagation mechanism of the oligomer formation [268]. Studies of the character of the kinetically controlled distribution function of macrocycles disclosed that there was a distinct correlation between concentrations of cyclics and the probability of ring closure. This is in obvious disagreement with any ring expansion mechanism. Thus it gives evidence that macrocycles are formed as result of the end-biting [268]. It is worth to point out that even large rings having more than sixty atoms in skeleton are formed in this way.

5.3.10 Propagation in Competition with Condensation

It is sometimes the case in cationic polymerization that growth of macromolecules occurs not only by the monomer addition to polymer chain, but also by condensation of end groups. The latter reaction reproduces directly or indirectly the initiator or active propagating centre and in that sense may be considered as a part of chain transfer process. This condensation, if occurs intramolecularly, leads to the formation of macrocycles. The simplest scheme of this mechanism is as follows:

$$\bigcirc + BA \longrightarrow B\sim\sim A$$

$$B(\sim\sim)_n A + M \longrightarrow B(\sim\sim)_{n+1} A \quad \text{(Growth by addition)}$$

$$B(\sim\sim)_n A + B(\sim\sim)_m A \longrightarrow B(\sim\sim)_{n+m} A + BA \quad \text{(Growth by condensation)}$$

$$B\sim\sim A \longrightarrow \bigcirc + BA \quad \text{(Cyclization)}$$

This kind of mechanism is observed in cationic polymerization of cyclosiloxanes[270]. It was shown that in the polymerization of hexamethylcyclotrisiloxane in the presence of CF_3SO_3H the monomer ist mostly consumed by the addition. However, the condensation has strong impact on the kinetics of this process[270].

5.3.11 Induction Periods in Polymerization Due to the End-Biting Reaction

In the polymerization of several cyclic acetals and ethers autoacceleration was observed. There might be a number of reasons for this behaviour, and some of these have already been discussed, namely the preinitiation equilibria and the inequality $k_i < k_p$. The enhanced reactivity of the hydroxy end group in polyacetals toward active species (in comparison with acetal groups) is another, not yet considered, reason for induction periods. When the polymerization degree increases with conversion the proportion of the active tertiary oxonium ions also increases, at the cost of the less reactive secondary oxonium ions (cf. Ref. 164), because the back-biting or intermolecular transfer become more important than the end-biting. Thus, there is no need to make speculative assumptions about the nature of the active species and to propose the "two stage polymerization of acetals"[272, 273] in order to explain the induction periods. Apparently, the autoacceleration observed in the polymerization of 1,3-dioxolane initiated by $HClO_4$[149], can also be explained by enhanced formation of the nonpropagating secondary oxonium ion at the early stages of polymerization. It is understandable, that in this case the depropagation-propagation cycles, even repeated many times, like in Plesch's experiments[149] give every time acceleration period.

5.3.12 Kinetic Depression of Macrocycles

This is a phenomenon opposite to that described in the previous section. A linear polymer is rapidly formed and then the equilibrium concentration of macrocycles is slowly reached by back-biting reactions according to the scheme

$$\sim\sim\oplus + M \xrightarrow[\text{fast}]{k_p} \sim\sim\oplus \; ; \; \sim\sim\oplus \underset{}{\overset{k_b(\text{slow})}{\rightleftharpoons}} \sim\sim\oplus + \bigcirc \qquad (166)$$

The kinetic depression is well known in the anionic polymerization of ε-caprolactone, where the formation of the cyclic dimer, trimer and tetramer starts when monomer is almost completely exhausted[274, 275].

In the cationic polymerization of heterocycles, a similar phenomenon was observed by Goethals in the polymerization of propylene sulfide and trans 2,3-dimethylthiirane[264]. The latter monomer polymerizes rapidly and quantitatively to a linear polymer which is then relatively slowly converted into 3,4,6,7-tetramethyl-1,2,5-trithiepane (167a). In this particular process, the macroring formation is a practically irreversible reaction and differs in this sense from the equilibrium processes discussed so far. The irreversibility is due to the formation of one molecule of cis-butene per one molecule of a cyclic trithiepane[276]:

$$\left[\text{S–CH–CH}\right]\cdots \longrightarrow \quad 167a \quad + \quad \text{H}_3\text{C–C=C–CH}_3 \quad (167)$$

167a

5.3.13 Stereochemistry of Back-Biting Processes. Stereoselectivity

There are three major factors governing the concentrations of macrocycles formed in back-biting reactions, namely ring strain, the probability of the ring to close, related to the Gaussian statistics (entropic factor), and the specific steric arrangement either facilitating or preventing a ring of a given size to be formed.

Polymerization of some heterocyclic monomers does not give the usual distribution of rings of various sizes but just one, specific macroring, sometimes in a high yield. Three-membered rings are usually converted to dimeric six-membered rings since these are least strained. A comprehensive list of these oligomers is published in Ref. 264, but it has to be remembered that probably all heterocycles produce cyclic oligomers, whereas their formation was studied only in some systems.

There are two reports clearly showing the overwhelming influence of steric factors on the selection of the macroring to be formed. Thus, ethylene oxide and its derivatives with small substituents give trimers or dimers and higher oligomers while ethylene oxides with large substituents (t-butyl[145], nitroethyl[277]) give almost exclusively tetramers. A similar general trend is observed for three-membered sulfides and amines.

Tsuruta[145] observed in the cyclooligomerization of t-butyloxirane that both (R)- and (S)-t-butyloxiranes predominantly yield cyclic tetramers; however, all the other macrocycles (from dimer to octamer) have also been observed in smaller concentrations:

$$4 \quad \triangle \underset{O}{} C_4H_9\text{-}t \quad \longrightarrow \quad$$

$$\begin{array}{ccc}
t\text{-}C_4H_9 & & C_4H_9\text{-}t \\
& CH\text{-}CH_2\text{-}O\text{-}CH & \\
& | & | \\
& O & CH_2 \\
& | & | \\
H_2C & & O \\
& | & | \\
& CH\text{-}O\text{-}CH_2\text{-}CH & \\
t\text{-}C_4H_9 & & C_4H_9\text{-}t
\end{array}$$

(168)

168a

The analysis of the ^{1}H- and ^{13}C-NMR spectra reveals that the cyclic tetramer of (R)-t-butyl-oxirane (*168a*) has exclusively gauche$^+$-gauche$^+$-trans (G$^+$G$^+$T) conformation. This indicates an exclusive attack on CH$_2$ causing inversion of configuration. On the other hand, in the linear poly-mer the G$^+$G$^-$T conformation of the main chain was deduced from NMR. From the analysis of the data reported in Ref. 145 it becomes clear that when the trimeric linear fragment of a macro-molecule is in the right position to form the G$^+$G$^+$T conformer cyclization occurs. However, when the conformation favors the G$^+$G$^-$T sequence, then the linear fragment is fixed in the growing macromolecule by the next addition. The conformation of the cyclic tetramer of (R)-t-butyloxirane is shown in Fig. 18. The yield of this tetramer is >65%.

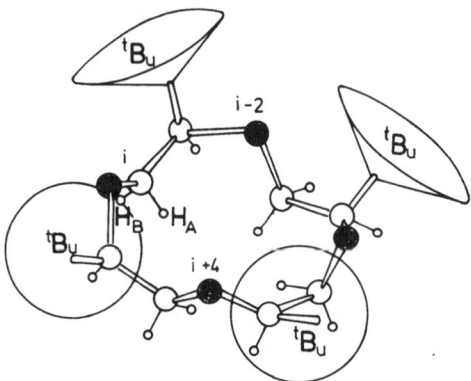

Fig. 18. Structure of the cyclic tetramer of (R)-t-butyloxirane. ◯: C; ●: O; ○: H; tBu: C(CH$_3$)$_3$ (Ref. 145)

This stereoselectivity in cyclooligomerization has not been observed for (RS)-t-butyloxirane. The presence of two enanthiomers, (R) and (S), apparently decreases the probability of the formation of the preferred conformation of the linear chain which leads to the G$^+$G$^+$T sequence.

Highly specific oligomerization was also observed for trans- and cis-1,2-dimethyl-thiirane.

Trans-1,2-dimethylthiirane gives a cyclic product, 1,2,5-trithiepane, which is formed in a kinetically distinct back-biting step when monomer is already converted into a linear polymer[276].

The trans monomer has the configuration (SS or RR). In the S$_N$2 attack in the propagation step, inversion of configuration on the carbon atom being attacked should take place (with all reservations discussed in Sect. 4.1) leading to a polymer with erythro configuration (RS or SR):

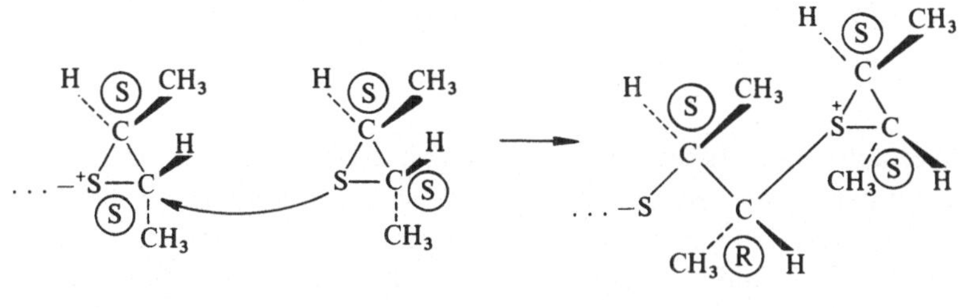

$$169a \qquad\qquad erythro \quad 169b \tag{169}$$

Due to the ambident reactivity of sulfur and the steric hindrance at the C atom, back-biting predominantly occurs as an attack of the S atom from the chain on the positively charged S atom (thiophilic attack) at the chain end with the elimination of an olefin:

$$170a \qquad\qquad 170b \qquad 170c \tag{170}$$

In the next step the three-membered cyclic sulfonium ion *170a* is regenerated by the back-biting attack of the penultimate unit (cf. discussion of the equilibrium between sulfonium ions in the polymerization of propylene sulfide on p. 109) followed by a thiophilic attack leading again to *170b* (a different mechanism was proposed in the original paper[276]). Comparison of this sequence of reactions with those proposed for propylene sulfide clearly indicates that the formation of one dominating cyclic oligomer is caused by steric hindrance (e.g. formation of trithiepane instead of a trimer in 1,2-dimethylthiirane cyclooligomerization).

It has conclusively been shown that, although six different isomers can be imagined, only two with specific stereochemistry are formed from trans-1,2-dimethylthiirane[276].

5.4 Hydride Ion, Proton and Carbocation Transfer to Monomer

Transfer and termination reactions may involve a number of different compounds present in the polymerization system; these are either added intentionally or present as impurities. The intentionally added compounds, usually chain transfer agents, are used simply either to decrease the molecular weight or to obtain polymers with controlled molecular weights and specific end groups (e.g. hydroxy and carboxy groups) which can then be utilized for further reactions.

In this section we shall, however, confine ourselves only to those reactions that involve monomer and polymer. The other transfer and termination reactions leading to oligomers of a desired molecular structure will be discussed in the second part of this review describing the synthetic applications of the cationic ring-opening polymerization.

5.4.1 Hydride Ion Shift and Transfer Reactions

According to well-established facts the hydride anion can only be shifted or transferred to the carbenium ion; the onium ions themselves are no hydride anion acceptors. Nevertheless, some of the onium ions exist in equilibrium with the corresponding carbenium ions (cf. Sect. 3.1) and in this isomeric form can abstract hydride anions. The classical experiments were performed by Jaacks in the polymerization of 1,3-dioxolane initiated by methoxymethylium perchlorate[68]:

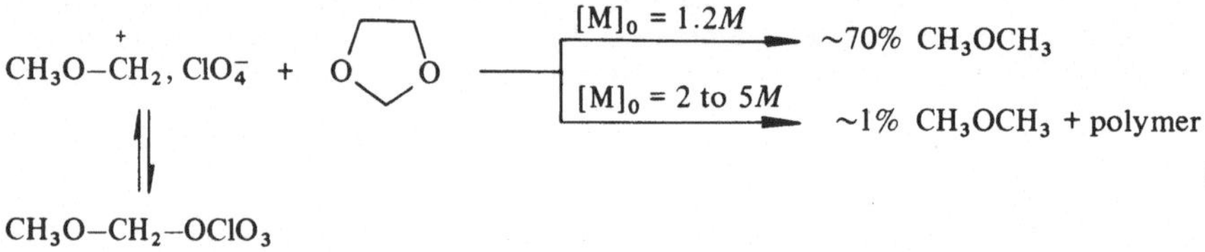

$$CH_3O{-}\overset{+}{C}H_2, ClO_4^- \;+\; \text{(dioxolane)} \longrightarrow \begin{cases} [M]_0 = 1.2M \longrightarrow \sim 70\% \; CH_3OCH_3 \\ [M]_0 = 2 \text{ to } 5M \longrightarrow \sim 1\% \; CH_3OCH_3 + polymer \end{cases}$$

$$\updownarrow$$

$$CH_3O{-}CH_2{-}OClO_3$$

$$(171)$$

The reaction between the methoxycarbenium ion and 1,3-dioxolane has already been discussed in Sect. 3.2. On the basis of these recent studies, the results quoted above can be explained by the following equilibria and competing reactions of chain transfer and chain propagation:

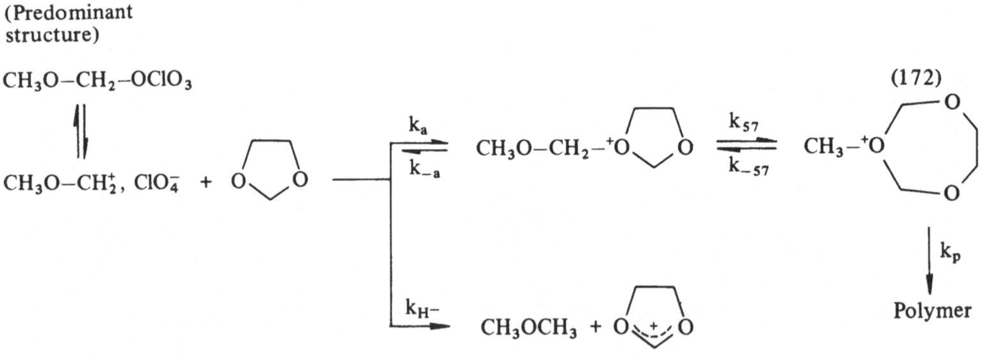

(Predominant structure)

$$CH_3O{-}CH_2{-}OClO_3 \qquad\qquad (172)$$

$$\updownarrow$$

$$CH_3O{-}\overset{+}{C}H_2, ClO_4^- + \text{(dioxolane)} \underset{k_{-a}}{\overset{k_a}{\rightleftharpoons}} CH_3O{-}CH_2{-}\overset{+}{O}\text{(ring)} \underset{k_{-57}}{\overset{k_{57}}{\rightleftharpoons}} CH_3{-}\overset{+}{O}\text{(ring)}$$

$$\downarrow k_p$$

$$\text{Polymer}$$

$$\overset{k_{H^-}}{\longrightarrow} CH_3OCH_3 + \text{(dioxolanylium)}$$

Thus, at lower $[DXL]_0$, the concentration of methoxycarbenium ions is higher than at higher $[DXL]_0$ because practically no polymerization occurs ($[DXL]_0 < [DXL]_e$). Eventually, all the initiator is converted into the dimethyl ether. Starting from sufficiently high $[DXL]_0$ the slow hydride transfer to the methoxycarbenium ion (k_{H^-} is probably much lower than both k_a and k_{57}, although it has not yet been measured) vanishes because the initiator is rapidly converted into the polymer end group.

At polymer-monomer equilibrium, there may still be a small concentration of polymeric alkoxycarbenium ions existing at equilibrium with the polymeric oxonium ions. Indeed, the hydride anion transfer was documented on the basis of the slowly increasing concentration of the $OHC-O-CH_2-\ldots$ polymer end groups observed directly by ^{1}H-NMR (the formic proton at 8.1 ppm). These groups are generated as a result of hydride anion transfer from monomer (better hydride donating agent) and polymer, e.g.:

(173)

$$\ldots-O-CH_2-CH_2-O-\overset{+}{C}H_2 \;+\; \underset{O\diagdown\diagup O}{\bigcirc} \longrightarrow \ldots-O-CH_2-CH_2-O-CH_3 \;+\; \underset{O\overset{+}{\smile}O}{\bigcirc}$$

followed by initiation of a new chain (cf. Sect. 3.2) by dioxolenium salt formed in the hydride transfer reaction

$$\underset{O\overset{+}{\smile}O}{\bigcirc} \;+\; \underset{O\diagdown\diagup O}{\bigcirc} \longrightarrow \underset{H}{\overset{O}{\diagdown}}C-O-CH_2-CH_2-\overset{+}{O}\underset{O}{\bigcirc} \xrightarrow{\;\text{Propagation}\;}$$

(174)

and, eventually, by insertion of the formic ester at the chain end.

    The hydride anion transfer reactions can almost be completely avoided in the polymerization of 1,3-dioxolane at lower temperature and polymers with polymerization degrees $\overline{DP}_n$ equal to $\overline{DP}_n$ (calc.)$= ([M]_0 - [M]_e)/[I]_0$ could be obtained (cf. Sect. 4.1). Model studies of the interaction between $CH_3O\overset{+}{C}H_2$ cation and cyclic and linear acetals have indeed shown that hydride anion transfer can hardly be observed below $-30\,°C$. It was also suggested that the original hydride transfer can be decreased if stronger nucleophiles (e.g. ethers) are added[51].

    Reliable evidence of hydride anion transfer also comes from the polymerization of 1,3,5-trioxane. Here, analytical studies have shown that polytrioxane contains approximately the same number of formyloxy and methoxy end groups[278]. Transacetalization (cf. Sect. 5.3) that accompanies polymerization ensures a random distribution of these end groups in the polymer molecules. Formation of these end groups from a polymer chain proceeds as follows:

$$\ldots-OCH_2OCH_2OCH_2OCH_2-\ldots \;+\; {}^+CH_2OCH_2OCH_2-\ldots \longrightarrow$$

$$\longrightarrow \ldots-OCH_2O\overset{+}{C}HOCH_2OCH_2-\ldots \;+\; CH_3OCH_2OCH_2-\ldots$$

(175)

$$\ldots-OCH_2O\overset{+}{C}HOCH_2OCH_2-\ldots \longrightarrow \ldots-OCH_2O\underset{H}{\overset{O}{C\diagup}} \;+\; \overset{+}{C}H_2OCH_2-\ldots \;\text{ etc.}$$

    Although the corresponding quantitative measurements of the carbenium-oxonium ion equilibrium in the polymerization of 1,3,5-trioxane have not yet been carried out, one can assume that the proportion of oxonium ions relative to this monomer will be lower than with 1,3-dioxolane which is much more nucleophilic. Moreover, the importance of hydride anion transfer may be higher for 1,3,5-trioxane polymerization. Thus, in spite of 1,3,5-trioxane itself being a much weaker hydride donating agent than 1,3-dioxolane, the hydride anion transfer processes play an important role in the polymerization of the former.

    There is also a possibility that intramolecular hydride shift takes place in the polymerization of 1,3,5-trioxane eventually leading to the same polymeric dialkoxycarbenium ion as shown in Eq. (175).

    In the polymerization of heterocyclic monomers other than cyclic acetals, there is no reliably documented cases of hydride anion transfer.

5.4.2 Proton and Carbocation Transfer Reactions

Proton transfer to monomer is the best known and most widely studied transfer reaction in the cationic polymerization of vinyl monomers. Cationic polymerization of styrene and isobutylene belong to the comprehensively studied examples[279]. More recently, it has been shown by ^{1}H-NMR that there are two kinds of double bonds formed as result of H^+ transfer in the polymerization of isobutylene[280]:

$$\ldots -CH_2-\underset{\underset{CH_3}{|}}{\overset{\overset{CH_3}{|}}{C}}-CH_2-C\!\!\begin{array}{l}\diagup CH_2 \\ \diagdown CH_3\end{array} \qquad\qquad \ldots -CH_2-\underset{\underset{CH_3}{|}}{\overset{\overset{CH_3}{|}}{C}}-CH=C\!\!\begin{array}{l}\diagup CH_3 \\ \diagdown CH_3\end{array}$$

Vinylidene type   Trisubstituted olefin type

Proton transfer in these reactions usually proceeds with higher activation enthalpy than the propagation reaction, giving the possibility of obtaining higher molecular weight polymers at lower temperatures.

Termination involving intramolecular proton shift or intermolecular proton transfer, both followed by elimination of a molecule of water has been reported for the high temperature cationic polymerization of lactams[31, 281]:

a) intramolecular:

$$\ldots -NH-\underset{}{\overset{\overset{O}{\|}}{C}}\ \ \overset{+}{N}H_3 \rightleftharpoons (\ldots -NH-\underset{}{\overset{\overset{OH}{|}}{C}}-\overset{+}{N}H_2) \rightleftharpoons \ldots -NH-\underset{}{\overset{\overset{H}{|}}{C}}\!\!=\!\!\overset{+}{N} + H_2O \qquad (176)$$

b) intermolecular involving a monomer molecule:

$$\ldots -\overset{+}{N}H_3 + CO-NH \rightleftharpoons \ldots -NH_2 + \underset{}{\overset{\overset{O}{\|}}{C}}-\overset{+}{N}H_2 \rightleftharpoons$$

$$\qquad\qquad\qquad\qquad\qquad\qquad\qquad\qquad\qquad\qquad\qquad\qquad (177)$$

$$\rightleftharpoons \ldots -NH-\underset{}{\overset{\overset{OH}{|}}{C}}-\overset{+}{N}H_2 \rightleftharpoons \ldots -NH-\underset{}{\overset{\overset{H}{|}}{C}}\!\!=\!\!\overset{+}{N} + H_2O$$

The detailed path of these reactions has not yet been established but the formation of cyclic amides at the chain ends was supported by their direct determination in macromolecules.

The amidine groups are more basic than the primary amino groups and decrease the proportion of protonated monomer. In this way, the concentration of the active species (proportional to the concentration of protonated monomer) is reduced thus decreasing the overall rate of polymerization.

In the cationic ring-opening polymerization of unsubstitued azetidine (178a) a proton is transferred from the dimeric ion (178c) to the monomer molecule. This process leads to unexpected structure of the polymeric backbone[282]:

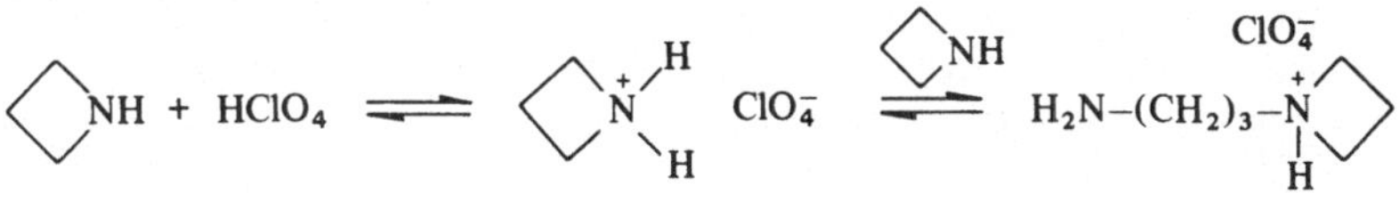

$$\text{(178)}$$

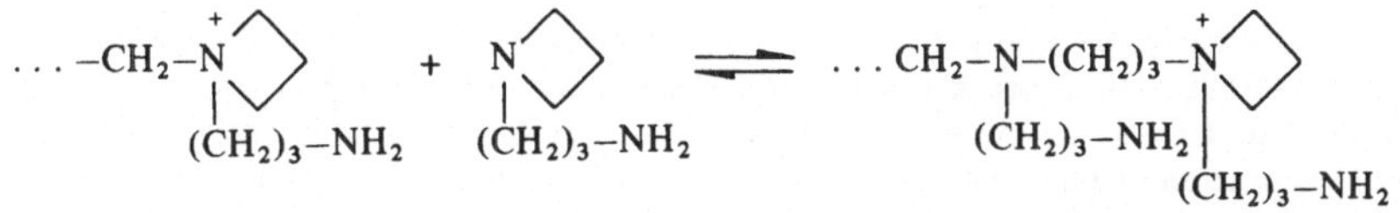

$$\text{(179)}$$

Carbocation transfer to monomer, closely related to proton transfer, has been reported for the polymerization of the six-membered cyclic esters of phosphoric acid[253]:

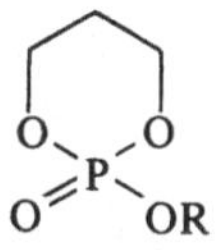

$R = CH_3$

The carbocation transfer has been deduced from the polymer structure studied by ^{1}H-NMR and ^{13}C-NMR. Polymers of molecular weights not exceeding 3000 have been obtained and the structure of their end groups can be best explained by carbocation transfer to monomer[253]:

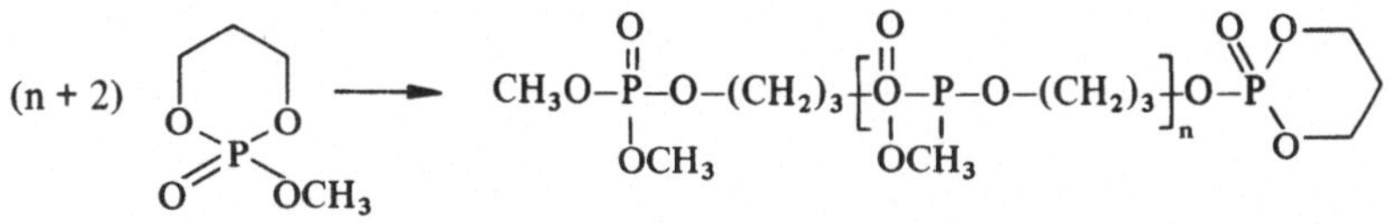

$$(n+2)\quad\xrightarrow{\qquad}\quad CH_3O-\overset{\overset{O}{\|}}{P}-O-(CH_2)_3\left[\overset{\overset{O}{\|}}{O}-P-O-(CH_2)_3\right]_n O-\overset{\overset{O}{\|}}{P}\qquad\text{(180)}$$

$^{31}P-NMR;\quad \delta = 6.0 \qquad\qquad -1.0 \qquad\qquad 0.0 \qquad\qquad 7.0\ ppm$

To explain the formation of oligomers having one cyclic end group, the following competition between chain growth and chain transfer was proposed:

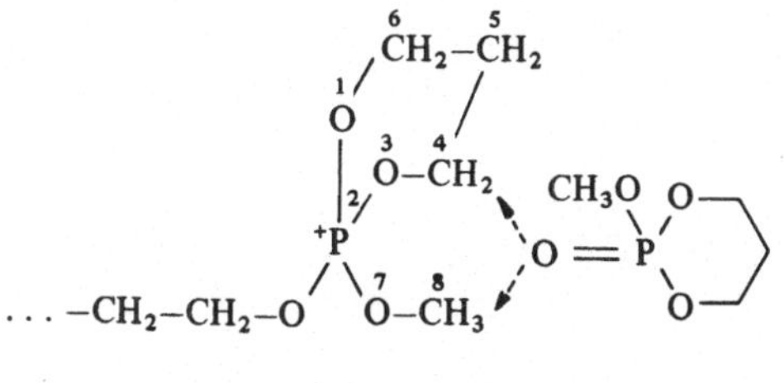

In tetraalkoxyphosphonium ions (shown above) a partial positive charge is localized at C-4 and C-6 (endocyclic C-atoms) as well as at exocyclic C-8. Therefore, when nucleophilic attack of the incoming monomer molecule takes place at C-4 (or C-6), chain propagation occurs. If, however, the reaction occurs at the exocyclic carbon atom (C-8), then the O-7−C-8 bond is cleaved, resulting in carbocation transfer to monomer:

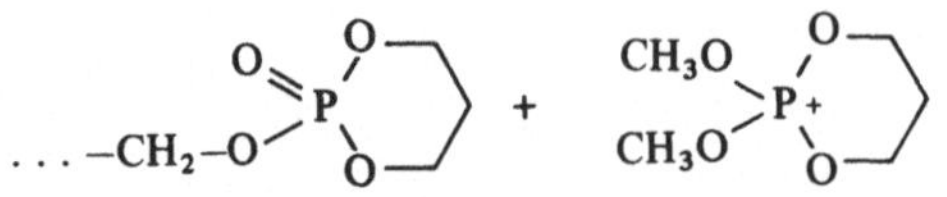

$$\ldots -CH_2-O \underset{\underset{O}{}}{\overset{O}{\underset{}{\overset{\displaystyle\nwarrow}{P}}}}{\Big\rangle} \quad + \quad CH_3O \underset{\underset{O}{}}{\overset{O}{\underset{}{\overset{\displaystyle/}{P+}}}}{\Big\rangle} \tag{181}$$

Because of the low ring strain in the six-membered ring of a cyclic phosphate (approx. $1.0 \, \text{kcal} \cdot \text{mole}^{-1}$) both reactions proceed with comparable rate constants ($\overline{DP}_n \leqslant 20$).

In this case, chain transfer has no kinetic effect because reinitiation is sufficiently fast.

In the polymerization of the higher 2-alkoxy-2-oxo-1,3,2-dioxaphosphorinanes, the combined GLC-MS method revealed that the following gaseous products are formed (exocyclic groups in the corresponding monomers are given first) ethyl: ethylene, n-propyl, i-propyl: propylene; cyclohexyl: cyclohexene. Apparently, in monomers with larger alkyl substituents, the formation of an olefin and proton transfer (in place of "CH_3^+" transfer observed for the 2-methoxy monomer) to the incoming monomer molecule proceed by a concerted mechanism.

Acknowledgements

This work was accomplished within a program 03.4 of the Polish Academy of Sciences.

One of us (S.P.) wishes to thank the Director and the Heads of the Chemical Laboratories of the Centre de Recherches sur les Macromolécules (CRM) in Strasbourg, France, for inviting him to CRM where a large part of this review has been prepared. Particular thanks are extended to Professor P. Rempp and Dr. E. Frante for their friendly help and numerous discussions.

Addendum

Because of the time gap between manuscript preparation and publication, coupled with a high activity in this field, we considered it desirable to update this review. Coverage of the relevant literature has been extended up to December 1979 and in a few cases to the early 1980. In some instances these new data required the older literature (not covered in the major text) to be quoted.

This Addendum is simply broken into three parts: initiation, propagation, and chain transfer and termination.

Initiation

As it has already been indicated in the main text, there is a number of initiators available that give a clean initiation, without side reactions, at least in case of the most thoroughly studied monomers, like cyclic acetals and cyclic ethers (mostly THF). Mono-, bi-, and multifunctional initiators were devised and anybody looking for an initiator for screening experiments has a choice of initiators (e.g. $(CF_3SO_2)_2O$ or PF_5) far superior to BF_3 or its complexes, the most widely used in the past. This superiority stems from the higher stability of anions derived from the former initiators. $\overline{BF}_4$ or $\overline{BF}_3OR$ anions are known to break easily and give highly nucleophilic $\overline{F}$ or $\overline{OR}$ anions causing termination by ions colapse.

Nevertheless, there is still an interest in finding either initiators emerging from new concepts of cation formation or in understanding initiation process for grafting from the backbone. In the later case these reactions can involve also chemical groups intentionally introduced into the backbones in order to get an efficient initiation.

To the first category belongs the photochemical formation of carbenium ions or protonic acids directly in the polymerization medium; this field, discussed in Sect. 3.1 and 3.2 has recently been reviewed by Smets at the IUPAC Macromolecular Symposium[283]. When cyclohexene oxide is used as a monomer the order of reactivities for iodonium or sulphonium salts, giving photochemically protonic acid, depend on the structure of anion MtX_n^- in the following way:

$$Ar_2I^+MtX_n^- \xrightarrow[\text{solv. RH}]{h\nu} ArI + ArH + Ar\text{-}Ar + HMtX_n$$

$$Ar_3S^+MtX_n^- \qquad Ar_2S + ArH + Ar\text{-}Ar + HMtX_n$$

$$SbF_6^- > AsF_6^- > PF_6^- > BF_4^- \qquad (182)$$

In the original paper[284] this order was explained by ion separation differences although it also coincides with anion stabilities (cf. Sect. 5.1.1).

Skorokhodov has recently used an unusual method for initiation. Although applied till now only for vinyl monomers, it is worth noticing, because of its potential use also for the cationic ring-opening polymerization. The method is based on the β-decay of tritium or organic tritium compounds[285].

$$^3H_2 \longrightarrow {}^3He{-}^3H^+ + \beta^- + \bar{\nu}; \quad {}^3He{-}^3H^+ \longrightarrow {}^3He + {}^3H^+$$
$$\text{or} \quad R{-}^3H \longrightarrow [R{-}^3He]^+ + \beta^- + \bar{\nu} \tag{183}$$
$$R^+ \overset{\downarrow}{+} He$$

High molecular weight polyisobutylene ($M_n = 1.2 \cdot 10^6$) was formed in 36% yield after 11 days at $-78°$ in bulk by using this method. Also polystyrene of an unusually high (for the cationic process) molecular weight ($M_n = 1.7 \cdot 10^6$) was obtained, however, in 4% yield only after 27 days at $0°$.

One of the most versatile groups of initiators, described in various sections of the review, is trifluoromethanesulfonic acid and its derivatives. Fujinaga and Sakamoto gave a detailed description of electrochemical characteristics of the acid itself and its derivatives in non-aqueous solvents[286]. The major properties of the acid, including the thermal and hydrolytic stability of the anion are described and the different orders of acid strengths, reported by various authors on the basis of conductometric measurements are given (e.g.: $CF_3SO_3H > HClO_4 > HBr > HJ > {} > FSO_3H > H_2SO_4 > HCl$ in acetic acid but $FSO_3H > ClSO_3H > CF_3SO_3H \gg HClO_4$ in sulfuric acid).

Perchloric acid and triflic acid behave as strong acids in basic solvents (like DMSO or DMF) but in solvents like acetonitrile the anions start to conjugate with the corresponding undissociated acids. This is because the anions compete with other nucleophiles in their reactions with acid. In order to understand better this competition the following mental experiment can be performed.

Let us first take a solution of e.g. triflic acid (HT) in CH_2Cl_2 and then add a nucleophile. In CH_2Cl_2 hydrogen-bonded associates of acid and free acid are involved in multiple equilibria. Ionization proceeds with formation of nucleophilic anion, which bonds by hydrogen bonding to the free acid (one or more molecules of acid are involved):

$$\text{e.g. } HT + HT \rightleftharpoons H^+, {}^-T \cdots HT \tag{184}$$

Addition to this system of a stronger nucleophile than anion itself leads to protonation of this nucleophile and formation of the hydrogen-bonded complexes:

$$H^+, T^- \cdots HT + Nu \rightleftharpoons H^+{-}Nu, T^- + HT \tag{185}$$

The position of equilibrium (185) will depend on the ratio of nucleophilicities of Nu and T^-, with strong nucleophiles in a highly ionizing media protonated nucleophile can be the only product.

This explanation is in a good agreement with differences observed when cationic polymerization of vinyl and heterocyclic monomers is initiated with triflic acid. In the former case some acid is bound to anions (e.g. 3 molecules of acid per anion[287]) whereas in the latter (e.g. 1,3-dioxolane)[34] every molecule of acid used gives one macromolecule.

The review covering the application of the acid, especially in organic chemistry has also been published[288].

Ziliox a.o. studied efficiency and mechanism of initiation of THF polymerization induced by several alkyl halides and AgSbF$_6$[289]. There are three different reaction paths considered by the authors:

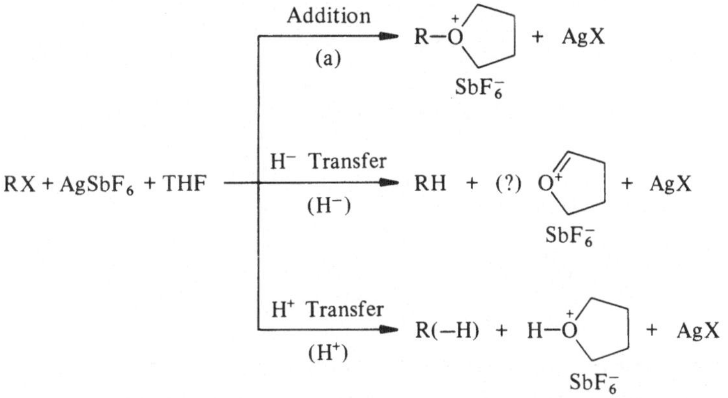

$$
RX + AgSbF_6 + THF \quad
\begin{cases}
\xrightarrow[\text{(a)}]{\text{Addition}} & R-\overset{+}{O}\!\!\diagdown\!\!\bigcirc + AgX \quad SbF_6^- \\[2ex]
\xrightarrow[\text{(H$^-$)}]{\text{H}^-\ \text{Transfer}} & RH + (?)\ O\overset{+}{\diagup}\!\!\bigcirc + AgX \quad SbF_6^- \\[2ex]
\xrightarrow[\text{(H$^+$)}]{\text{H}^+\ \text{Transfer}} & R(-H) + H-\overset{+}{O}\!\!\diagdown\!\!\bigcirc + AgX \quad SbF_6^-
\end{cases}
\quad (186)
$$

Below, in Table 19 these results are summarized and all of the three reactions paths indicated with the letters referring to scheme (186).

Table 19. Reaction path in the system RX + THF + AgSbF$_6$ (Scheme 186)[289]

Alkyl halide (corresponding to cation)	Halogen atom		
	I	Br	Cl
$(C_6H_5)_3C^+$	H^-	H^-	H^-
$(C_6H_5)_2CH^+$			a
$C_6H_5CH_2^+$		a	H^+
p-$CH_3C_6H_4CH_2^+$		a^a	
$(CH_3)_3C^+$	H^+	H^+	H^+
$(CH_3)_2CH^+$	a	H^+	
$CH_2=CH-CH_2^+$		a/H^+	

^{a} Ref. 290, AgPF$_6$ salt

Table 19 has been compiled by using data taken mostly from Ref. 289.

It would be probably instructive to take two examples from Table 19 in order to put a rationale for these data. There is a competition, in every case, between addition, H^-, and H^+ transfer. The competition between addition and H^- transfer is discussed in Section 3.2.3 with triphenylmethylium cation taken as an example (no H^+ expulsion possible for structural reasons).

Let us consider the second group: addition vs H^+ transfer. Trimethylmethylium and dimethylmethylium cations constitute good example. The former initiates by H^+ transfer and the latter by direct addition. One could propose a scheme formally similar to the addition/ H^- competition:

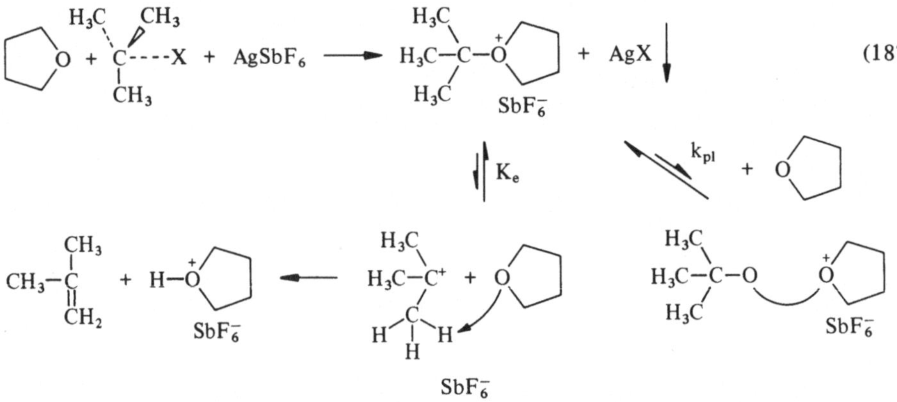

(187)

Although K_e and k_{pl} were not measured but they both should be higher for di-methylmethylium cation than for trimethylmethylium cation. Thus, higher concentration of oxonium ion (which, as we think, do not transfer a proton by itself) and higher rate of addition of the THF molecule to the first oxonium ion eliminate the possibility of proton transfer. In all of the studied till now cases, whenever the H^+ transfer has structurally been possible, this reaction dominated over the H^- transfer (cf. Scheme (186)).

Direct initiation with metal (and metalloid) halides, described in Sect. 3.2.9 for THF-PF_5 system, has recently been studied by Vladimirova a.o.[291] for α-epichloro-hydrin and $SnCl_4$. The original proposal of Eastham[292], who advanced the dizwitter-ionic growth for oxirane and $SnCl_4$, has been confirmed by using the new analytical methods: ^{14}C tracer and phenoxy end-capping.

Recently, hydronium ion was observed directly by ^{17}O-NMR and sp^2 hybridization of oxygen was suggested[293].

Propagation

Stereocontrol in Polymerization

In 1977 Deslongchamps formulated, on the basis of a number of observations of different reactions (e.g. ozonolysis, hydrolysis) of heterocyclic compounds the following rule of reactivity[294]:

"whenever two heteroatoms and a leaving group are linked to the same carbon atom, specific cleavage of a carbon-oxygen bond in any conformer is allowed only if the other two heteroatoms each have an orbital oriented antiperiplanar to the leaving 0-alkyl group".

This rule, born as a purely experimental observation has been applied very recently to explain some phenomena in the polymerization of cyclic amides by Bertalan[295] and cyclic orthoesters by Hall[296].

An alternative mechanism of the cationic polymerization of lactams was postulated by Bertalan[295]. The protonated amino end-group existing in equilibrium with a tetrahedral intermediate (188a) was assumed as the growing species.

In order to explain the termination reaction, leading to the cyclic amidate cation through removal of the water molecule, Bertalan proposed the antiperiplanar participation (structure 188b):

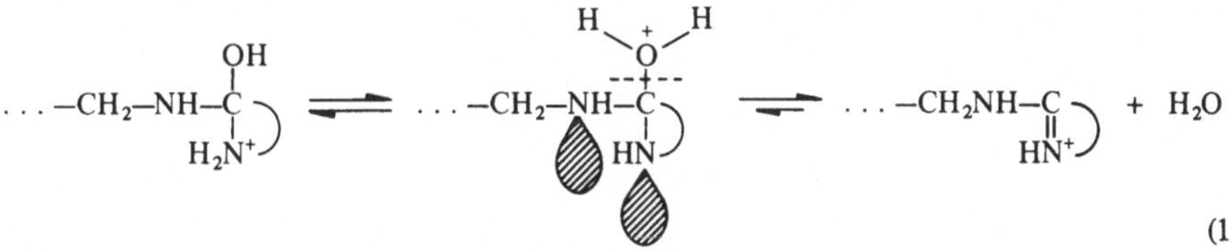

(188)

188a   188b

It seems to us however that propagation, proceeding via tetrahedral intermediate should involve activated monomer. Nucleophilic attack of the amine on positively charged carbon atom in activated monomer molecule is followed by proton transfer either to oxygen or nitrogen atoms. The cleavage of C–O bond (termination) in the intermediate is accelerated by two antiperiplanar orbitals at N atoms (188b). C–N bond cleavage (propagation) is favoured by the ring strain and one antiperiplanar orbital at oxygen atom. The exocyclic C–N bond breaks in the depropagation step:

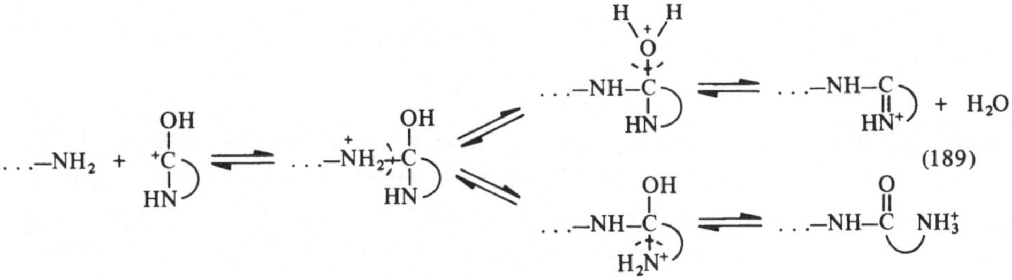

(189)

Hall, in the polymerization of the bicyclic orthoester:

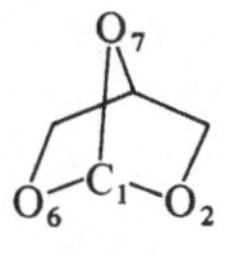

190

2,6,7-Trioxabicyclo[2,2,1]heptane

observed, that at low temperature predominantly polymers containing five-membered dioxolane rings in the chain are formed:

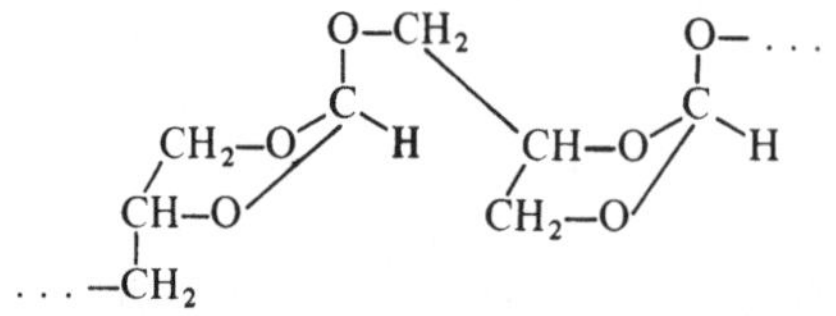

$$
\begin{array}{c}
\text{O–CH}_2 \qquad\qquad \text{O–} \ldots \\
\text{CH}_2\text{–O} \quad\text{C} \quad\text{H} \qquad \text{CH–O} \quad\text{C} \quad\text{H} \\
\text{CH–O} \qquad\qquad \text{CH}_2\text{–O} \\
\ldots\text{–CH}_2 \\
\mathit{191}
\end{array}
$$

Formation of this structure requires breaking of the $C_1–O_2$ or $C_1–O_6$ bond in *190*.

At higher temperature the six-membered rings became apparent as well in the polymer chain:

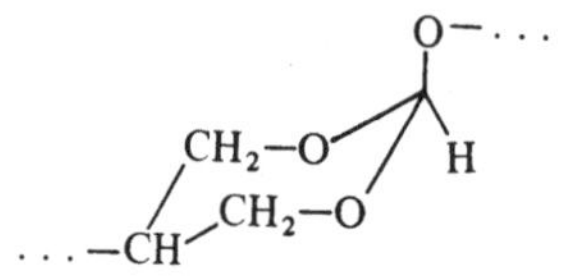

$$
\begin{array}{c}
\text{O–} \ldots \\
\text{CH}_2\text{–O} \qquad \text{H} \\
\text{CH}_2\text{–O} \\
\ldots\text{–CH} \\
\mathit{192}
\end{array}
$$

i.e. the bond $C_1–O_7$, within a five-membered ring (cf. *190*) starts to break more often.

The proportions of structures *191* and *192* were found from the relative peak areas at δ 5.24 and δ 5.81, assigned to the orthoformate protons as shown above.

Comparison of the structures of oxonium ions leading to either *191* or *192*:

e. g. 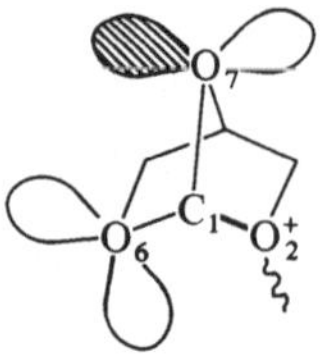 and

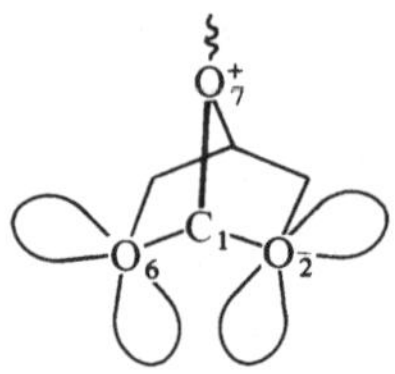

194

(solid lines show the bonds to be broken in order to obtain the five-membered rings in the chain (193 → 191) or six-membered ones (194 → 192)) shows that in order to obtain polymer with 5-membered rings in the chain one has to break 6-membered ring, whereas for obtaining a polymer with 6-membered rings in the chain breaking of 5-membered ring is required. Thus, remembering that the difference in the ring strains between 5- and 6-membered rings is approx. 4 kcal · mol^{-1} one has to look for an explanation why the less strained ring in the bicyclic monomer is broken. We shall follow the Hall's reasoning in this respect. First of all, the corresponding instantaneous concentration of two rival growing species (*193* and *194*) can be different; O_2 (or O_6) oxygen atoms are more nucleophilic than O_7 and, thus *193* will prevail, leading to the kinetically controlled in this case polymer structure *191*. Another possibility comes from the already discussed stereoelectronic control. The $C_1–O_2$ bond in *193* is in the antiperiplanar position to one lone electron pair at O_7

whereas in 194 C_1-O_7 bond is handicapped by not being exposed to any (neither at O_2 nor at O_6) electron pair in antiperiplanar position.

Although, according to the original paper by Deslongchamps the stereoelectronic control requires assistance of two neighbouring antiperiplanar electron pairs, but more recently Kirby has shown that even one electron pair is sufficient to accelerate cleavage of an acetal bond in antiperiplanar position[297, 298]. Thus, it is not clear whether the described above kinetic control or stereocontrol allow to open predominantly the less strained ring or whether both effects work hand in hand overshadowing the differences in ring strain.

Comparison of ring strains requires one more comment. The quoted above difference of 4 kcal $\cdot$ mole^{-1} between the strains of 1,3-dioxolane and 1,3-dioxane refers to the neutral molecules whereas on competition between active species *193* and *194* the corresponding oxonium ions are involved. In our knowledge there are, however, no data available on the ring strains of cyclic oxonium ions.

Okada[299] polymerized 6,8-dioxabicyclo[3,2,1]octane:

(195)

and observed that the polymer obtained at low temperature with $BF_3 \cdot O(C_2H_5)_2$ initiator from racemic monomer has ^{13}C-NMR spectrum in which all of the signals are compatible with a fully "α-form" of a polymer:

(196)

(α-form in terminology of carbohydrate chemistry means that the exocyclic oxygen atom is in the axial position to the ring).

Signals are split into two closely located singlets and comparison of the ^{13}C-NMR spectrum of the racemic polymer, obtained from the racemic monomer, with that of the optically active polymer (prepared from $(+)$-(1 R, 5 S)-6,8-dioxabicyclo[3,2,1]octane indicate, that the lower field, smaller peak of each signal pair comes from the D-L (syndiotactic) dyad. The higher field signal can thus be ascribed to the dyad structures of D-D and L-L consecutive units (isotactic dyads).

At thigher temperature of polymerization the β-structure appears at the spectra and polymerization looses its stereospecificity. This can be due to some participation of carbenium ions; their proportion should increase with temperature (cf. Sec. 3.2.2).

Preferential formation of the isotactic dyads (L-L and D-D units), i.e. the higher rates of homopropagation than of crosspropagation has been explained by Okada in terms of stereoselection due to the steric hindrance when, e.g. the D-monomer approaches the growing center composed of L-unit:

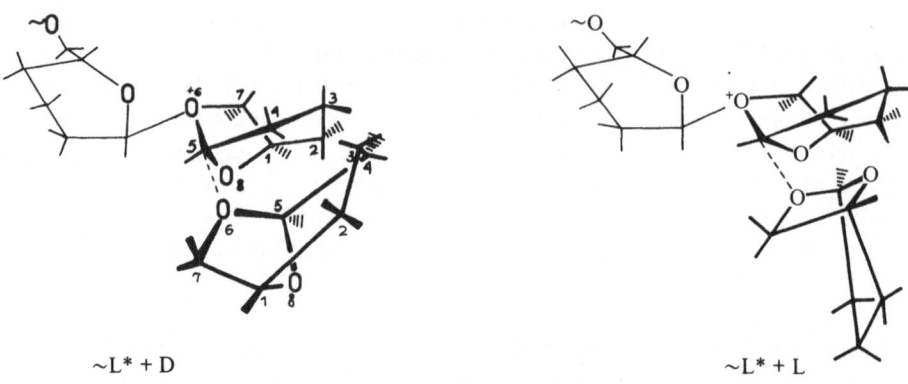

~L* + D   ~L* + L

Fig. 19. Possible scheme for polymerization of 6,8-dioxabicyclo I 3,2,1 I octane[299]

The clear cut case of enantiomorphic selection has been reported by Vanderberg[300] in polymerization of trans-2,3-epoxybutanes:

$$
\begin{array}{cc}
\text{(structure 1)} & \text{(structure 2)}
\end{array}
\tag{197}
$$

Average ratio of triads RS-RS-RS to RS-RS-SR was found to be 1.4. No syndiotactic triads RS-SR-RS were detected. Thus, polymer consists of blocks made by 5 consecutive additions of one stereoisomer, followed by one addition of the second stereoisomer. Formation of these sequences and absence of the syndio-triads confirms again the propagation with inversion, as required for the S_N2 addition to the oxonium ion. Crystalline poly(cis-2,3-epoxybutane) was confirmed (cf. Sect. 4.4) to have predominantly racemic RR-RR-RR (or SS-SS-SS) triads. It is also interesting to note that the racemic polymer with regular triads RR-RR-RR was observed from meso (RS) oxide. These findings confirm the Vanderberg's earlier proposal[301] that the ring-opening polymerization proceeds with inversion of configuration at carbon atom.

In the studies of microstructure of poly(cyclohexene oxide) obtained with $(C_6H_5)_3C^+AsF_6^-$ initiator Blanchard a.o. observed the mm, mr and rr triads. The intensity values are equal to 0.15, 0.47 and 0.39 respectively[302]. The obtained polymers are amorphous, both cis- and trans-structures are present.

Since it has been demonstrated, however[215], that in S_N1 mechanism, proceeding with participation of the ion-pairs, some inversion of configuration is possible (earlier reserved exclusively for S_N2 processes involving oxonium ions) the stereochemistry of the products lost its diagnostic power for reaction mechanism. Thus, it would be premature to draw conclusions about the structures of growing species on the basis of data given by Blanchard[302]. Nevertheless, the high probability of racemic diads formation ($P_r = 0.62$) may indicate participation of carbenium ions.

Termination and Transfer Processes

Macrocyclization. End-Biting and Back-Biting

The importance of simultaneous end-biting and back-biting, according to the general mechanism of propagation accompanied by transfer due to cyclization[156, 266]:

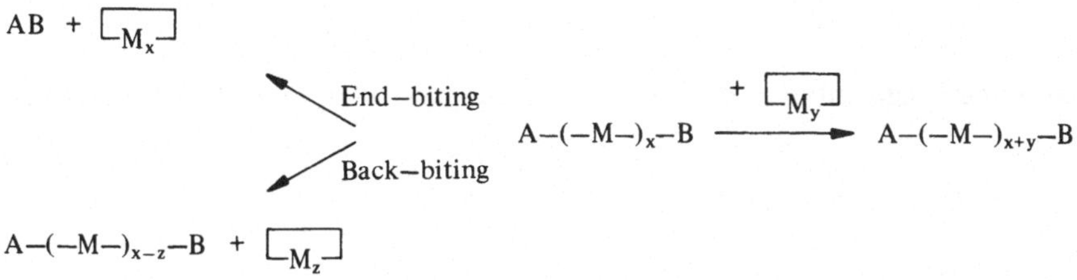

$$AB \ + \ \boxed{-M_x-}$$

End—biting

Back—biting

$$A-(-M-)_x-B \xrightarrow{+ \ \boxed{-M_y-}} A-(-M-)_{x+y}-B$$

$$A-(-M-)_{x-z}-B \ + \ \boxed{-M_z-} \tag{198}$$

has recently been shown in the polymerization of ethylene oxide and its copolymerization with THF[303]. A clear distinction was made for polymerization with non-hydrolizable protonic acids (like CF_3SO_3H furnishing A = OH in a scheme above) and initiators providing A being an ether end-group. Only macrocyclic ethers could have been detected by mass spectrometry for polymerizations induced by acids.

In another series of papers, published by Yamashita a.o. and describing macrocyclization in the cationic polymerization of oxiranes, much more complicated mixture of cyclic products was observed. Besides typical macrocyclic polyethers[304-306] cyclic oligomers having acetal structure were found. These products can be formed, according to Yamashita, as a result of back-biting to the rearranged growing species:

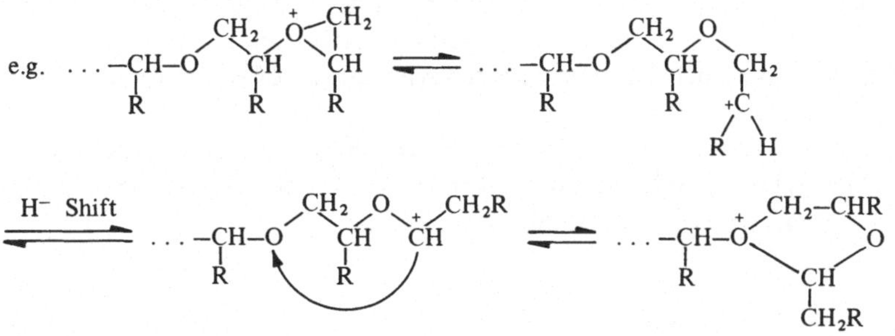

$$\tag{199}$$

The rearrangement, as postulated by Yamashita, seems probable, because of the formation of the carbenium ion stabilized by adjacent oxygen atom.

Cyclooligomerization of substituted oxiranes has been found to be highly stereospecific[304]. Cis-2,3-dimethyloxirane gave only one dioxolane-type dimer namely 2,cis-4,trans-5-trimethyl-2-ethyl-1,3-dioxolane isomer (plus other cyclic- and linear oligomers):

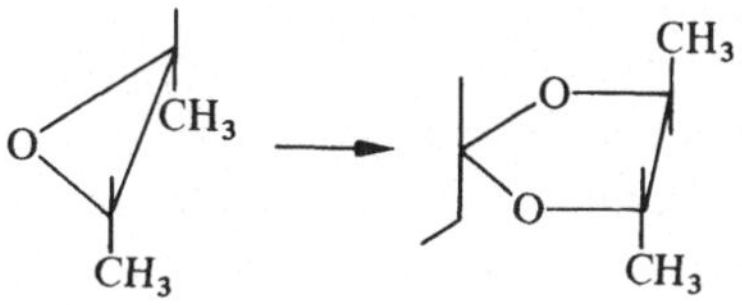

$$\tag{200}$$

Similar "isomerised" dimers were observed in the polymerization of 2,2-di-methyl-[305] and 2-phenyloxiranes[306], whereas in the polymerization of 2-chloro-methyloxirane[307] only 1,4-dioxane type dimer was obtained. These results are compatible with the explanation, requiring the formation of carbenium ion; the presence of electron-donating methyl or phenyl groups made apparently this rearrangement possible. Higher yield of rearranged species was also observed in solvents of higher polarity, stabilizing the carbenium ion formed.

Termination by Fragmentation of Complex Anions and Collapse of the Pairs of Ions

Studies pertinent to this termination:

$$\ldots -\overset{+}{O}\langle\text{ring}\rangle \rightleftharpoons \ldots -O \diagdown X + MtX_n \tag{201}$$

$$MtX_{n+1}^-$$

have been performed by two groups[308, 309].

Jones and Plesch[308] have reported on decomposition of triethyloxonium salts with BF_4^-, PF_6^-, and SbF_6^- in CH_2Cl_2 solution. At $60°$ rate constants of decomposition were as follows (in s^{-1}): $3.3 \cdot 10^{-4}$ (BF_4^-), $7 \cdot 10^{-5}$ (PF_6^-), $1.3 \cdot 10^{-6}$ (SbF_6^-). Thus SbF_6^- anion is at these conditions approx. 50 times more stable than PF_6^- anion. Decomposition rate is influenced by additions of nucleophiles into reaction system (e.g. $(C_2H_5)_2O$, 1,3-dioxolane). Solvation shell, formed by additives, retards oxonium ion against attack of the anion, because overall enthalpy of activation increases upon addition of ether ($\Delta\Delta H^{\ddagger} = 7$ kcal $\cdot$ mole^{-1}). Entropy of activation increases even to higher extent ($\Delta\Delta S^{\ddagger} \leqslant 30$ e.u.). This is connected with release of solvating ether molecules in the course of reaction.

Heublein a.o.[309] have studied stability of various complex anions in decomposition of triphenylmethylium salts, assuming that the rate constant of collapse of ion pair can be used as a measure of anion stability:

$$(C_6H_5)_3C^+MtX_{n+1}^- \overset{k}{\longrightarrow} (C_6H_5)_3CX + MtX_n \tag{202}$$

This reaction was studied conductometrically and by UV-spectroscopy. Once again the most stable anion was SbF_6^-. It was surprising that in trityl salts PF_6^- anion has been found to be 10^3 times less stable than BF_4^- and 10^2 times less stable than $SbCl_6^-$ anion, on the contrary to anion stabilities found for oxonium salts. Substitution of one halogen atom by alkoxy or hydroxy group makes no significant difference in anion stability e.g. $SbCl_6^-$ ($k = 3 \cdot 10^{-5} s^{-1}$), $SbCl_5OH^-$ ($k = 2 \cdot 10^{-5} s^{-1}$), $SbCl_5OCH_3^-$ ($k = 2 \cdot 10^{-5} s^{-1}$). Stability of anions with mixed ligands (e.g. $SbCl_3Br_3^-$) is additive function of content of more tightly bound (smaller) halogen atoms.

Termination due to the collapse of the growing macroion-pair was detected by ^{1}H-NMR in oligomerization of 2-alkoxy-1,3,2-dioxaphospholanes by Yamashita a.o.[310]. Oligomers of phosphonate structure were formed in reaction and simultaneously transfer occurred as it could be expected for Arbusov rearrangement:

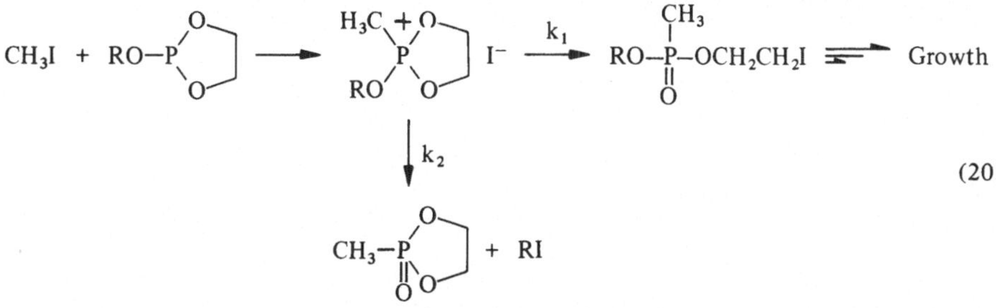

$$CH_3I + RO-P\underset{O}{\overset{O}{\diagdown}}\longrightarrow \underset{RO}{\overset{H_3C}{\diagdown}}\overset{+}{\underset{O}{P}}\overset{O}{\diagdown} I^- \xrightarrow{k_1} \underset{\overset{\|}{O}}{RO-\overset{CH_3}{\underset{|}{P}}-OCH_2CH_2I} \rightleftharpoons Growth$$

$$\Big\downarrow k_2$$

$$CH_3-\underset{\overset{\|}{O}}{P}\overset{O}{\diagdown} + RI$$

(203)

Enthalpies of activation of both reactions, proceeding with rate constants k_1 and k_2 are nearly the same for 2-methoxy substituted monomer. Ratio $k_1/k_2 = 3$ was calculated in $40°$ and $100°$. 2-Neopenthoxy substituted monomer does not participate in Arbuzov type transfer and the rate constant k_1 is approximately 40 times lower than for 2-methoxy derivative. When substituent was changed to t-buthoxy group decomposition of intermediate salt into isobutylene, 2-hydro-2-oxo-1,3,2-dioxaphospholane and 2-methyl-2-oxo-1,3,2-dioxaphospholane was observed.

Transfer to Polymer

Transfer to polymer (Sect. 5.3) has been described by Aleksiuk a.o.[311] in polymerization of 3-chloromethyl-3-methyloxetane initiated with aluminum alkyls. High molecular weight polymers ($M_n = 6$ to $8 \cdot 10^5$) with narrow molecular weight distribution ($M_w/M_n = 1.29$) were formed from very beginning of reaction and were independent of conversion. These data suggest slow initiation, fast propagation and unimolecular decomposition process. Authors propose the termination on the own polymer backbone and formation of nonreactive branched oxonium ion. Similar mechanism described earlier for 3,3-bis(chloromethyl)oxetane is discussed in Sect. 3.2.11.

In the polymerization of 1-oxa-3-thiacyclopentane:

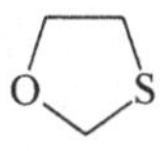

$$\underset{O\diagdown\diagup S}{\bigcirc}$$

(204)

only limited yields of low-molecular weight polymers were obtained[312]. This fact was attributed to the known higher nucleophilicity of linear sulfides than of their cyclic counterparts, leading to the fast conversion of the reactive, cyclic sulfonium cations to the branched ones, as discussed above for oxetanes:

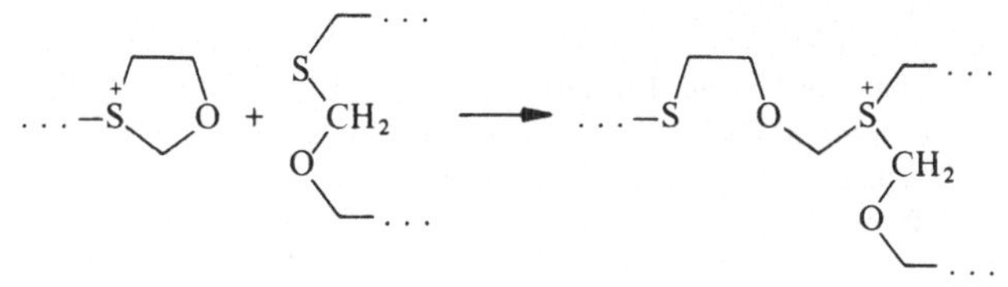

$$\ldots-\overset{+}{S}\underset{O}{\diagdown}O + \underset{O}{\overset{S}{\diagdown}}CH_2 \longrightarrow \ldots-S\underset{O}{\diagdown}O\underset{CH_2}{\overset{+}{S}}\cdots$$

(205)

References

1. Plesch, P. H. (Ed.): The Chemistry of Cationic Polymerization. Oxford: Pergamon Press 1963
2. Kennedy, J. P.: Cationic Polymerization of Olefins. A Critical Inventory. New York: Wiley Interscience 1975
3. Pepper, D. C., Reilly, P. J.: Proc. Roy. Soc. Ser. A 291, 41 (1966)
4. Dainton, F. S., Sutherland, G. B. B. M.: J. Polym. Sci. 4, 37 (1949)
5. Bawn, C. E. H. et al.: Polymer 12, 119 (1971)
6. Sawamoto, M., Higashimura, T.: Polym. Prep. 20, 727 (1979)
7. Sawamoto, M., Higashimura, T.: Macromolecules 11, 328 (1978)
8. Szwarc, M.: Carbanions Living Polymers and Electron Transfer Processes. New York: Interscience 1968
9. Schulz, G. V.: Chem. Techn., 220 (1973)
10. Bywater, S.: Adv. Polym. Sci. 4, 66 (1965)
11. Sigwalt, P., Boileau, S.: J. Polym. Sci., Polym. Symp. 62, 51 (1978)
12. Kazanskij, K. S.: Khimija i Technologija Vysokomol. Soed. 9, 5 (1977)
13. Penczek, S., Matyjaszewski, K.: J. Polym. Sci., Polym. Symp. 56, 255 (1976)
14. Pell, A. S., Pilcher, G.: Trans. Faraday Soc. 61, 71 (1965)
15. Ivin, K. J.: In: Polymer Handbook, 2nd Ed., Brandrup, J.' Immergut, E. H. (Eds.) New York: Wiley 1975, p. II-421
16. Busfield, W. K., Lee, R. M., Merigold, D.: Makromol. Chem. 156, 183 (1972)
17. Lambert, J. B. et al.: J. Amer. Chem. Soc. 96, 6112 (1974)
18. Romers, C. et al.: In: Topics in Stereochemistry, Eliel, E. L., Allinger, N. L. (Eds.) New York: Wiley Interscience 1969, Vol. 4, p. 39
19. Arnett, E. M.: In: Progress in Physical Organic Chemistry, New York: Interscience 1963, Vol. I., p. 243
20. Searles, Jr. S., Tamres, M.: Basicity and complexing ability of ethers. In: The Chemistry of the Ether Linkage. Patai, S. (Ed.). London, New York, Sydney: Wiley 1967, p. 243
21. Kabir-ud -Din, Plesch, P. H.: J. Electroanal. Chem. 93, 29 (1978)
22. Chmelir, M., Cardona, N., Schulz, G. V.: Makromol. Chem. 178, 169 (1977)
23. Sauvet, G., Vairon, J. P., Sigwalt, P.: J. Polym. Sci., Polym. Symp. 52, 173 (1975)
24. Bell, R. P.: The Proton in Chemistry. Ithaca, New York: Cornell Univ. Press 1973 p. 22
25. Shchori, E., Jagur-Grodzinski, J.: J. Amer. Chem. Soc. 94, 7957 (1972)
26. Klages, F., Meuresch, H., Steppich, W.: Ann. Chem. Liebiegs 592, 81 (1955)
27. Meerwein, H. et al.: J. Prakt. Chem. 154, 83 (1939)
28. Jones, F. R., Plesch, P. H.: Chem. Commun. 1969, 1231
29. Brouwer, D. M.: Tetrahedron Letters 1968, 453
30. Wurtz, A.: Compt. rend. 50, 1197 (1860)
31. Šebenda, J.: Lactams. In: Comprehensive Chemical Kinetics. Bamford, C. H., Tipper, C. F. H. (Eds.). Vol. 15, pp. 379–471. Amsterdam, Oxford, New York: Elsevier 1976
32. Rothe, M., Boenisch, H., Kern, W.: Makromol. Chem. 67, 90 (1963)
33. Pruckmayr, G., Wu, T. K.: Macromolecules 6, 33 (1973)
34. Kubisa, P., Penczek, S.: in preparation

35. Eigen, M.: Pure & Appl. Chem. *6*, 97 (1963)
36. Kunitake, T., Takarabe, K.: J. Polym. Sci., Polym. Symp. *56*, 33 (1976)
37. Kunitake, T., Takarabe, K.: Polym. J. *10*, 105 (1978)
38. Souverain, D. PhD Thesis (P. Sigwalt, Thesis adviser) Univ. Paris VI, 1978
39. Morozova, I. S. et al.: Dokl. Akad. Nauk SSSR *209*, 153 (1973)
40. Rasvodovskii, E. F. et al.: J. Macromol. Sci. A *8*, 241 (1974)
41. Van Ooteghem, D., Goethals, E. J.: Makromol. Chem. *177*, 3389 (1976)
42. Woodhouse, M. E., Lewis, F. P., Marks, T. J.: J. Amer. Chem. Soc. *100*, 996 (1978)
43. Ledwith, A.: Adv. Chem. Ser. *91*, 317 (1969)
44. Sauvet, G., Vairon, J. P., Sigwalt, P.: J. Polym. Sci. A 1 *7*, 983 (1969)
45. Gogolczyk, W., Slomkowski, S., Penczek, S.: J. Chem. Soc. Perkin II *1977*, 1729
46. Burns, F. W. et al.: IUPAC Congress, Helsinki 1972, I-30
47. Kalfoglou, N., Szwarc, M.: J. Phys. Chem. *72*, 2233 (1968)
48. Bowyer, P. M., Ledwith, A., Sherrington, D. C.: J. Chem. Soc. (B) *1971*, 1511
49. Cotrel, R. et al.: Macromolecules *9*, 931 (1976)
50. Dreyfuss, M. P., Westfahl, J. C., Dreyfuss, P.: Macromolecules *1*, 437 (1968)
51. Penczek, S.: Makromol. Chem. *175*, 1217 (1974)
52. Kuntz, I., Melchior, M. T.: J. Polym. Sci. A *7*, 1959 (1969)
53. Kubisa, P., Penczek, S.: Makromol. Chem. *144*, 169 (1971)
54. Afshar-Taromi, F. et al.: Makromol. Chem. *179*, 849 (1978)
55. Slomkowski, S., Penczek, S.: J. Chem. Soc. Perkin II, *1974*, 1718
56. Meerwein, H. et al.: Ann. Chem. Liebiegs *635*, 1 (1960)
57. Pittman, Jr., C. U., McManus, S. P., Larsen, J. W.: Chem. Rev. *72*, 357 (1972)
58. Berlin, Al. Al. et al.: Vysokomol. Soed. *12*, 443 (1970)
59. Slomkowski, S., Penczek, S.: Chem. Commun. *1970*, 1347
60. Kabir-ud -Din, Plesch, P. H.: J. Chem. Soc., Perkin II *1978*, 937
61. Jedliński, Z. et al.: Macromolecules *9*, 622 (1976)
62. Slomkowski, S.: PhD Thesis, Lodz, 1974
63. Stolarczyk, A., Kubisa, P., Penczek, S.: Bull. Acad. Pol. Sci. Ser. Sci. Chem. *22*, 431 (1974)
64. Pihlaja, K., Äyräs, P.: Acta. Chem. Scand.: *24*, 531 (1970)
65. Olah, G. A., Svoboda, J. J.: Synthesis *1973*, 52
66. Szymanski, R.: in preparation
67. Jaacks, V. et al.: Makromol. Chem. *115*, 290 (1968)
68. Boehlke, K., Weyland, P., Jaacks, V.: XXIII IUPAC Congr., Boston 1971, Vol. II p. 641
69. Perst, M.: Oxonium Ions in Organic Chemistry. The Hague: Verlag Chemie, Academic 1971
70. Kubisa, P.: Bull. Acad. Polon. Sci., Ser. Sci. Chem. *25*, 627 (1977)
71. Stolarczyk, A., Kubisa, P., Penczek, S.: J. Macromol. Sci.-Chem., A *11*, 2047 (1977)
72. Stolarczyk, A. et al.: XXIII IUPAC Congr., Boston, 1971, Vol. I, p. 178
73. Yamashita, Y. et al.: J. Polym. Sci. B *8*, 481 (1970)
74. Jedliński, Z. J., Łukaszczyk, J.: Macromolecules *8*, 700 (1975)
75. Olah, G. A. et al.: J. Amer. Chem. Soc. *84*, 2733 (1962)
76. Nuyken, O., Plesch, P. H.: Chem. and Ind. *1973*, 379
77. Meerwein, H., Delfs, D., Morschel, H.: Angew. Chem. *72*, 927 (1960)
78. Yamashita, Y. et al.: Makromol. Chem. *142*, 171 (1971)
79. Franta, E. et al.: J. Polym. Sci., Polym. Symp. *56*, 139 (1976)
80. Kubisa, P., Penczek, S.: Makromol. Chem. *179*, 445 (1978)
81. Kubisa, P.: J. Macromol. Sci.-Chem. A *11*, 2247 (1977)
82. Meerwein, H.: In: Houben-Weyl Methoden der Organischen Chemie. Müller, E. (Ed.), 4th edn. Vol. VI/3, Stuttgart: Georg Thieme Verlag, 1965 p. 325
83. Klages, F., Meuresch, H.: Chem. Ber. *85*, 863 (1952) and *86*, 1322 (1953)
84. Olah, G. A., Olah, J. A., Svoboda, J. J.: Synthesis *1973*, 490
85. Szymański, R. et al.: Chem. Commun. *1976*, 33
86. Ref. 82, p. 359
87. Saegusa, T. et al.: Macromolecules *6*, 657 (1973)
88. Matyjaszewski, K.: to be published

89. Drijvers, W., Goethals, E. J.: Makromol. Chem. *148*, 311 (1971)
90. Łapienis, G.: PhD Thesis, Łódź 1976
91. Nysenko, Z. N. et al.: Vysokomol. Soed. *18*, 1696 (1976)
92. Kubisa, P., Penczek, S.: Macromolecules *10*, *1216* (1977)
93. Jones, F. R., Plesch, P. H.: Chem. Commun. *1969*, 1230
94. Goethals, E. J. et al.: J. Macromol. Sci.-Chem. *A 7*, 1375 (1973)
95. Goethals, E. J. et al.: ACS Symposium Ser. *59*, 1 (1977)
96. Dafforn, G. A., Streitwieser Jr., A.: Tetrahedron Lett. *36*, 3159 (1970)
97. Streitwieser Jr., A.: Chem. Rev. *56*, 571 (1956)
98. Matyjaszewski, K.: PhD Thesis, Łódź 1976
99. Saegusa, T., Kobayashi, S., Yamada, A.: Makromol. Chem. *177*, 2271 (1976)
100. Smith, S., Hubin, A. J.: J. Macromol. Sci. *A 7*, 1399 (1973)
101. Smith, S., Schultz, W. J., Newmark, R. A.: ACS Symposium Ser. *59*, 13 (1977)
102. Marek, M., Chmelíř, M.: J. Polym. Sci. *C23*, 223 (1968)
103. McLaughlin, D. E., Tamres, M.: J. Amer. Chem. Soc. *82*, 5618 (1960)
104. Brown, H. C., Adams, R. M.: J. Amer. Chem. Soc. *64*, 2557 (1942)
105. Rutenberg, A. C., Palko, A. A.: J. Phys. Chem. *69*, 527 (1965)
106. Brownstein, S., Eastham, A. M., Latremouille, G. A.: J. Phys. Chem. *67*, 1028 (1963)
107. Fratiello, A., Onak, T. P., Schuster, R. E.: J. Amer. Chem. Soc. *90*, 1194 (1968)
108. Fratiello, A., Vidulich, G. A., Schuster, R. E.: J. Inorg. Nucl. Chem. *36*, 93 (1974)
109. Hoene, R., Reichert, K. H. W.: Makromol. Chem. *177*, 3545 (1976)
110. Andruzzi, F., Prescia, A., Ceccarelli, G.: Makromol. Chem. *176*, 977 (1975)
111. Entelis, S. G., Korovina, G. V.: Makromol. Chem. *175*, 1253 (1974)
112. Komratov, G. N., Barzykina, R. A., Korovina, G. V.: Vysokomol. Soed. *20*, 608 (1978)
113. Entelis, S. G.: private communication, Ufa, May 1979
114. Muetterties, E. L. et al.: J. Inorg. Nucl. Chem. *16*, 52 (1960)
115. Rose, J. B.: J. Chem. Soc. *1956*, 542
116. Farthing, A. C.: J. Chem. Soc. *1955*, 3648
117. Saegusa, T., Matsumoto, S., Hashimoto, Y.: Polymer J. *1*, 31 (1970)
118. Ledwith, A.: Polymer *19*, 1217 (1978)
119. Abdul-Rasoul, F. A. M., Ledwith, A., Yagci, Y.: Polymer *19*, 1219 (1978)
120. Crivello, J. V., Lam, J. H. W.: Macromolecules *10*, 1307 (1977)
121. Crivello, J. V., Lam, J. H. W.: Polym. Prepr. *20*, 415 (1979)
122. Penczek, S., Kubisa, P.: Makromol. Chem. *130*, 186 (1969)
123. Aleksiuk, G. P., Alferova, L. V., Kropachev, V. A.: Vysokomol. Soed., in press
124. Beste, L. F., Hall, Jr., H. K.: J. Phys. Chem. *68*, 269 (1964)
125. Matyjaszewski, K., Kubisa, P., Penczek, S.: J. Polym. Sci., Polym. Chem. Ed. *13*, 763 (1975)
126. Kobayashi, S., Danda, H., Saegusa, T.: Macromolecules *7*, 415 (1974)
127. Kobayashi, S., Morikawa, K., Saegusa, T.: Macromolecules *8*, 386 (1975)
128. Vofsi, D., Tobolsky, A. V.: J. Polym. Sci. *A 3*, 3261 (1965)
129. Brzezińska, K.: unpublished results
130. Kobayashi, S., Morikawa, K., Saegusa, T.: Macromolecules *8*, 952 (1975)
131. Kobayashi, S., Danda, H., Saegusa, T.: Bull. Chem. Soc. Jap. *47*, 2699 (1974)
132. Penczek, S. et al.: Makromol. Chem. *172*, 243 (1973)
133. Weissermel, K., Nölken, E.: Makromol. Chem. *68*, 140 (1963)
134. Sims, D.: J. Chem. Soc., 864 (1964)
135. Rosenberg, B. A. et al.: Vysokomol. Soed. *6*, 2030 (1964)
136. Bawn, C. E. H., Bell, R. M., Ledwith, A.: Polymer *6*, 95 (1965)
137. Dreyfuss, M. P., Dreyfuss, P.: Polymer *6*, 93 (1965)
138. Matyjaszewski, K., Penczek, S.: J. Polym. Sci., Polym. Chem. Ed. *12*, 1905 (1974)
139. Łapienis, G., Penczek, S.: Macromolecules *7*, 166 (1974)
140. Goethals, E. J., Drijvers, W.: Makromol. Chem. *165*, 329 (1973)
141. Goethals, E. J., Schacht, E. H.: J. Polym. Sci., Polym. Letters Ed. *11*, 497 (1973)
142. Kennedy, J. P., Johnston, J. E.: Adv. Polym. Sci. *19*, 57 (1975)

143. Rosenberg, B. A. et al.: Vysokomol. Soed. 6, 2035 (1964)
144. Estrin, J. I., Entelis, S. G.: Zhurnal Fiz. Chimii 43, 2837 (1969)
145. Sato, A.: Polymer J. 9, 209 (1977)
146. Matsuzaki, K., Ito, H.: J. Polym. Sci., Polym. Chem. Ed. 15, 647 (1977)
147. Vandenberg, E. J.: Pure & Appl. Chem. 48, 295 (1976)
148. Saegusa, T. et al.: Macromolecules 5, 233 (1972)
149. Plesch, P. H., Westermann, P. H.: J. Polym. Sci. C16, 3837 (1968)
150. Plesch, P. H.: Br. Polym. J. 5, 1 (1973)
151. Plesch, P. H.: Pure & Appl. Chem. 48, 287 (1976)
152. Boehlke, K., Jaacks, V.: Makromol. Chem. 145, 219 (1971)
153. Yokoyama, Y., Okada, M., Sumimoto, H.: Makromol. Chem. 178, 529 (1977)
154. Rosenberg, B. A., Irzak, W. I., Enikolopyan, N. S.: Interchain Exchange Reactions in Polymers (Russian). Moscow: Khimia 1975
155. Enikolopyan, N. S. et al.: Vysokomol. Soed. 19, 1924 (1977)
156. Penczek, S., Kubisa, P.: ACS Symposium Ser. 59, 60 (1977)
157. Brzeziñska, K. et al.: Makromol. Chem. 178, 2491 (1977)
158. Kubisa, P., Penczek, S.: Makromol. Chem. 180, 1821 (1979)
159. Yokoyama, Y., Okada, M., Sumimoto, H.: Makromol. Chem. 179, 1393 (1978)
160. Berman, E. L. et al.: Dokl. Akad. Nauk. SSSR 231, 1388 (1976)
161. Rothe, M., Bertalan, G.: ACS Symposium Ser. 59, 129 (1977)
162. Müller, E.: Neuere Anschauungen der Organischen Chemie. Berlin: Springer Verlag 1957
163. Lambert, J. B., Johnson, D. H.: J. Amer. Chem. Soc. 90, 1349 (1968)
164. Frenking, G., Kato, H., Fukui, K.: Bull. Soc. Chem. Jap. 48, 6 (1975)
165. Eizner, Yu. E., Erusalimskij, B. L.: Electronic Aspect of Polymerization (Russian). Leningrad: Nauka 1976
166. Anet, F. A. L., Osyany, J. M.: J. Amer. Chem. Soc. 89, 352 (1967)
167. Naumov, V. A.: Dokl. Akad. Nauk SSSR 169, 839 (1966)
168. Darwish, D., Tourigny, G.: J. Amer. Chem. Soc. 88, 4303 (1966)
169. Saegusa, T., Matsumoto, S.: J. Polym. Sci. A-1, 6, 1559 (1968)
170. Saegusa, T., Kobayashi, S.: In: Progress in Polymer Science, Japan 6, 107 (1973)
171. Barzykina, R. A. et al.: Vysokomol. Soed. 16, 906 (1974)
172. Matyjaszewski, K., Słomkowski, S., Penczek, S.: J. Polym. Sci., Polym. Chem. Ed. 17, 69 (1979)
173. Matyjaszewski, K., Słomkowski, S., Penczek, S.: J. Polym. Sci., Polym. Chem. Ed. 17, 2413 (1979)
174. Brzeziñska, K., Matyjaszewski, K., Penczek, S.: Makromol. Chem. 179, 2387 (1978)
175. Sherrington, D. C.: J. Polym. Sci., Polym. Symp. 56, 323 (1976)
176. Ma, C. C. et al.: J. Macromol. Sci.-Chem. A11, 1613 (1977)
177. Denison, J. T., Ramsey, J. B.: J. Amer. Chem. Soc. 77, 2615 (1955)
178. Ref. 8, p. 241
179. Yeager, H. L., Kratochvil, B.: J. Phys. Chem. 73, 1963 (1969)
180. Ledwith, A., Sherrington, D. C.: Adv. Polym. Sci. 19, 1 (1975)
181. Yao, N. P., Benion, D. N.: J. Electrochem. Soc. 118, 1097 (1971)
182. Dreyfuss, P., Dreyfuss, M. P.: Adv. Chem. Ser. 91, 335 (1969)
183. Goethals, E. J.: Makromol. Chem. 175, 1309 (1974)
184. Ref. 8, p. 411
185. Sangster, J. M., Worsfold, D. J.: Macromolecules 5, 229 (1972)
186. Bourdauducq, P., Worsfold, D. J.: Macromolecules 8, 562 (1975)
187. Bruggeman, P., Goethals, E. J.: 1st Internat. Symp. Polymerization of Heterocycles, Jablonna 1975, p. 111
188. Lyudvig, E. B., Khomyakov, A. K., Sanina, G. S.: J. Polym. Sci., Polym. Symp. 42, 289 (1973)
189. Belenkaya, B. G., Levenko, A. I., Lyudvig, E. B.: Vysokomol. Soed. 20, 559 (1978)
190. Dzavadyan, E. A., Rosenberg, B. A., Enikolopyan, N. S.: Vysokomol. Soed. 15, 1317 (1973)
191. Komarov, B. A. et al.: Vysokomol. Soed. A16, 2464 (1974)

192. Szwarc, M.: Adv. Chem. Ser. *91*, 236 (1969)
193. Ivin, K. J.: In: Reactivity, Mechanism and Structure in Polymer Chemistry. Jenkins, A. D., Ledwith, A. (Eds). London: Wiley 1974
194. Dincer, S., Van Ness, H. C.: J. Chem. Eng. Data *16*, 378 (1971)
195. Kuzub, L. I. et al.: Vysokomol. Soed. *10*, 2007 (1968)
196. Szwarc, M.: Macromolecules *11*, 1053 (1978)
197. Alder, R. W., Baker, R., Brown, J. M.: Mechanism in Organic Chemistry, p. 49. London: Wiley 1971
198. Saegusa, T., Matsumoto, S.: Macromolecules *1*, 442 (1968)
199. Saegusa, T. et al.: Macromolecules *5*, 34 (1972)
200. Saegusa, T. et al.: Macromolecules *6*, 26 (1973)
201. Saegusa, T. et al.: Macromolesules *5*, 815 (1972)
202. Saegusa, T. et al.: Macromolecules *5*, 236 (1972)
203. Schacht, E. H., Goethals, E. J.: Makromol. Chem. *167*, 155 (1973)
204. Schacht, E. H., Bossaer, P. K., Goethals, E. J.: Polymer J. *9*, 329 (1977)
205. Goethals, E. J.: J. Polym. Sci., Polym. Symp. *56*, 271 (1976)
206. Kagiya, T., Matsuda, T., Hirota, R.: J. Macromol. Sci.-Chem. *A 6*, 451 (1972)
207. Saegusa, T.: Makromol. Chem. *175*, 1199 (1974)
208. Plesch, P. H.: Adv. Polym. Sci. *8*, 137 (1971)
209. Brzezińska, K., Matyjaszewski, K., Penczek, S.: submitted for publication
210. Vandenberg, E. J.: J. Polym. Sci. *47*, 489 (1960)
211. Yokoyama, M. et al.: Macromolecules *5*, 690 (1972)
212. Zachoval, J., Schuerch, C.: J. Amer. Chem. Soc. *91*, 1165 (1969)
213. Hall, Jr., H. K. et al.: J. Amer. Chem. Soc. *96*, 7265 (1974)
214. Okada, M., Sumimoto, H., Hibino, Y.: Polym. J. *6*, 256 (1974)
215. Parker, R. E., Isaaks, N. S.: Chem. Rev. *59*, 737 (1959)
216. Kagiya, T. et al.: J. Macromol. Sci.-Chem. *A 6*, 1631 (1972)
217. Gumbs, R. et al.: Macromolecules *2*, 77 (1969)
218. Gandini, A., Plesch, P. H.: Eur. Polym. J. *4*, 55 (1968)
219. Matyjaszewski, K., Kubisa, P., Penczek, S.: Internat. Symp. Cationic Polymerization, Rouen France, Sept. 1973, paper C-25
220. Kobayashi, S., Danda, H., Saegusa, T.: Bull. Chem. Soc. Jap. *46*, 3214 (1973)
221. Wu, T. K., Pruckmayr, G.: Macromolecules *8*, 77 (1975)
222. Pruckmayr, G., Wu, T. K.: Macromolecules *8*, 954 (1975)
223. Kagiya, T., Matsuda, T.: J. Macromol. Sci.-Chem. *A 5*, 1265 (1971)
224. Matyjaszewski, K., Diem, T., Penczek, S.: Makromol. Chem. *180*, 1817 (1979)
225. Saegusa, T., Kobayashi, S.: J. Polym. Sci. Polym. Symp. *56*, 241 (1976)
226. Matyjaszewski, K., Zielinski, M.: J. Macromol. Sci.-Chem., *A 13*, 193 (1979)
227. Leonard, J., Maheux, D.: J. Macromol. Sci., A7, 1421 (1973).
228. Matyjaszewski, K., Penczek, S.: J. Polym. Sci., Polym. Chem. Ed. *15*, 247 (1977)
229. Buyle, A. M., Matyjaszewski, K., Penczek, S.: Macromolecules *10*, 269 (1977)
230. Matyjaszewski, K., Franta, E., Penczek, S.: Polymer *20*, 1184 (1979)
231. Winstein, S. et al.: Organic Reaction Mechanism Spec. Publ. *19*, 109 (1965)
232. Saegusa, T.: J. Macromol. Sci.-Chem. *A 6*, 997 (1972)
233. Volkov, V. P. et al.: Vysokomol. Soed. *16*, 2190 (1974)
234. Lyudvig, E. B.: IUPAC Internat. Symp. Macromolecular Chemistry, Budapest, Hungary 1969, paper 2−04
235. Penczek, S., Kubisa, P.: Makromol. Chem. *165*, 121 (1973)
236. Ivanov, V. V. et al.: Vysokomol. Soed. *16*, 243 (1973)
237. Burton, R. E., Pepper, D. C.: Proc. Roy. Soc. *A 263*, 58 (1961)
238. Penczek, S.: Bull. Acad. Pol. Sci., ser. sci. chim. *18*, 39 (1970)
239. Mateva, R., Kabaivanov, W.: Makromol. Chem. *178*, 2609 (1977)
240. Worsfold, D. J., Eastham, A. M.: J. Amer. Chem. Soc. *79*, 900 (1957)
241. Rose, J. B.: J. Chem. Soc. *1956*, 546
242. Dreyfuss, P., Dreyfuss, M. P.: Polym. J. *8*, 81 (1976)

243. Black, P. E., Worsfold, D. J.: Can. J. Chem. *54*, 3325 (1976)
244. Bucquoye, M., Goethals, E. J.: Makromol. Chem. *179*, 1681 (1978)
245. Stern, M. D., Tobolsky, A. V.: J. Chem. Phys. *14*, 93 (1946)
246. Weissermel, K. et al.: Kunststoffe *54*, 410 (1964)
247. Jaacks, V.: Makromol. Chem. *101*, 33 (1967)
248. Fejgin, J., Penczek, S.: J. Polym. Sci. *B 4*, 615 (1966)
249. Jaacks, V.: Adv. Chem. Ser. *91*, 371 (1969)
250. Rosenberg, B. A., Efremova, A. I., Enikolopyan, N. S.: Vysokomol. Soed. *7*, 2172 (1965)
251. Simonds, R. P., Goethals, E. J., Spassky, N.: Makromol. Chem. *179*, 1851 (1978)
252. Penczek, S.: Bull. Acad. Pol. Sci., ser. sci. chim. *18*, 53 (1970)
253. Łapienis, G., Penczek, S.: Macromolecules *10*, 1301 (1977)
254. Penczek, S.: Makromol. Chem. *134*, 299 (1970)
255. Jaacks, V., Kern, W.: Makromol. Chem. *83*, 71 (1965)
256. Vasiliev, N. I. et al.: Dokl. Akad. Nauk SSSR *176*, 831 (1967)
257. McKenna, J. M., Wu, T. K., Pruckmayr, G.: Macromolecules *10*, 877 (1977)
258. Rosenberg, B. A. et al.: Vysokomol. Soed. *7*, 188 (1965)
259. Franta, E., Afshar-Taromi, F., Rempp, P.: Makromol. Chem. *177*, 2191 (1976)
260. Vasiliev, N. I. et al.: J. Polym. Sci. *C22*, 679 (1969)
261. Taganov, N. G. et al.: Vysokomol. Soed. *19* B, 510 (1977)
262. Unger, S. H., Hansch, C.: In: Progress in Physical Organic Chemistry. Taft, R. W. (Ed.) Vol. *12*, 91 (1976)
263. Semlyen, J. A.: Adv. Pol. Sci. *21*, 41 (1976)
264. Goethals, E. J.: Adv. Pol. Sci. *23*, 103 (1977)
265. Jacobson, H., Stockmayer, W. H.: J. Chem. Phys. *18*, 1600 (1950)
266. Matyjaszewski, K. et al.: Makromol. Chem., *181*, 1469 (1980)
267. Schulz, R. C. et al.: ACS Symposium Ser. *59*, 77 (1977)
268. Chojnowski, J., Scibiorek, M., Kowalski, J.: Makromol. Chem. *178*, 1351 (1977)
269. Pruckmayr, G., Wu, T. K.: Macromolecules *11*, 265 (1978)
270. Chojnowski, J., Wilczek, L.: Makromol. Chem. *180*, 117 (1979)
271. Burwell, Jr., R. L., Fuller, M. E.: J. Amer. Chem. Soc. *79*, 2332 (1957)
272. Yamashita, Y., Kawakami, Y.: ACS Symposium Ser. *59*, 99 (1977)
273. Kawakami, Y., Suzuki, J., Yamashita, Y.: Polymer J. *9*, 519 (1977)
274. Ito, K., Yamashita, Y.: Macromolecules *11*, 68 (1978)
275. Yamashita, Y.: Polym. Preprints *20*, 126 (1979)
276. Van Craeynest, W., Goethals, E. J.: Eur. Polym. J. *12*, 859 (1976)
277. Entelis, S. G., Korovina, G. V., Kuzayev, A. I.: Vysokomol. Soed. *A 13*, 1438 (1971)
278. Weissermel, K. et al.: Angew. Chem. *79*, 512 (1967)
279. Kämmerer, H., Sayed-Mozaffari, A.: Internat. Symp. Cationic Polymerization, Rouen, France 1973, paper C-19
280. Puskas, I., Banas, E. M., Nerheim, A. G.: J. Polym. Sci. Polym. Symp. *56*, 191 (1976)
281. Puffr, R., Šebenda, J.: J. Polym. Sci., Polym. Chem. Ed. *12*, 21 (1974)
282. Goethals, E. J., Schacht, E. H.: 23. Internat. Symp. Macromolecules, Madrid, Spain 1974, paper I 3–26
283. Smets, G.: Pure & Appl. Chem. *52*, 253 (1980)
284. Crivello, J. V., Lam, J. H. W.: J. Polym. Sci., Chem. Ed. *17*, 977 (1979)
285. Akulov, G. et al.: Makromol. Chem. *179*, 2775 (1978)
286. Fujinaga, T., Sakamoto, I.: Pure & Appl. Chem. *52*, 1389 (1980)
287. Sigwalt, P., Sauvet, G.: 5th International Symposium on Cationic and other Ionic Polymerizations, Kyoto, Japan, 1980, p. 68
288. McCown, J. D.: Chem. Rev. *77*, 69 (1977)
289. Zilliox, J. G. et al.: 26th IUPAC symposium on Macromolecules, Mainz, FRG, 1979, Vol. I, p. 56
290. Richards, D. H., Thompson, R. D.: Polymer *20*, 1439 (1979)
291. Vladimirova, L. G. et al.: Vysokomol. Soed. *20 B*, 200 (1978)

292. Eastham, A.: in "The Chemistry of Cationic Polymerization". Plesch, P. H., Ed., Pergamon Press, Oxford, 1963
293. Matescu, G. D., Benedikt, G. M.: J. Amer. Chem. Soc. *101,* 3959 (1979)
294. Deslongchamps, P.: Tetrahedron *31,* 2463 (1975)
295. Bertalan, G. et al.: 26th IUPAC Symposium on Macromolecules, Mainz, FRG, 1979, Vol. I, p. 177
296. Yokoyama, Y. et al.: Macromolecules *13,* 252 (1980)
297. Kirby, A. J., Martin, R. J.: Chem. Commun. *1978,* 803
298. Kirby, A. J., Martin, R. J.: Chem. Commun. *1979,* 1079
299. Okada, M., Sumimoto, H., Komada, H.: Macromolecules *12,* 395 (1979)
300. Ho, F. F. L., Vanderberg, E. J.: Macromolecules *12,* 212 (1979)
301. Vanderberg, E. J.: Pure & Appl. Chem. *48,* 295 (1976)
302. Malhotra, S. L., Blanchard, L. P.: J. Macromol. Sci.-Chem. *A 12,* 1379 (1978)
303. Robinson, I. M., Pruckmayr, G.: Macromolecules *12,* 1043 (1979)
304. Kawakami, Y., Ogawa, A., Yamashita, Y.: J. Polymer Sci., Polym. Chem. Ed. *17,* 3785 (1979)
305. Yamashita, Y., Ito, K.: Pol. Bull. *1,* 73 (1978)
306. Yamashita, Y., Iwao, K., Ito, K.: J. Polymer Sci., Polym. Letters *17,* 1 (1979)
307. Ito, K., Usami, N., Yamashita, Y.: Polymer J. *11,* 171 (1979)
308. Jones, F. R., Plesch, P. H.: J. Chem. Soc. Dalton *1979,* 927
309. Heublein, G., Spange, S., Hallpop, P.: Makromol. Chem. *180,* 1935 (1979)
310. Kawakami, Y., Miyota, K., Yamashita, Y.: Polymer J. *11,* 175 (1975)
311. Aleksiuk, G. P., Alferowa, L. V., Kropachev, V. A.: Vysokomol. Soed. *21,* 2759 (1979)
312. Kawakami, Y., Mizutai, Y., Yamashita, Y.: Makromol. Chem. *180,* 2279 (1979)

T. Saegusa (editor)

Subject Index

Author Index Volumes 1–37

Polymer Bulletin

Editors:
Prof. H.-J. Cantow, Makromolekulare Chemie, Universität Freiburg, Stefan-Meier-Strasse 31, D-7800 Freiburg, West-Germany
Prof. J. P. Kennedy, Dept. of Polymer Science, The University of Akron, Akron, OH 44325, USA
Prof. T. Saegusa, Dept. Synthetic Chemistry, Kyoto University, Kyoto, 606, Japan

Editorial Board: H. Batzer, Basel; N. Calderon, Akron, OH; S. Cesca, San Donato Milanese; P. J. Flory, Stanford, CA; J. Furukawa, Tokyo; J. E. McGrath, Blacksburg, VA; H. K. Hall, Jr., Tucson, AZ; H. H. Kausch, Lausanne; T. Kelen, Budapest; M. Kryszewski, Lódź; A. Ledwith, Liverpool; E. Maréchal, Paris-Cedex; J. Meißner, Zürich; A. Nakajima, Kyoto; G. and S. Henrici Olivé, Research Triangle Park, NC; N. A. Plate, Moscow; B. Rånby, Stockholm; C. I. Simionescu, Bucureşti; S. Sivaram, Gujarat; D. H. Solomon, Melbourne; R. Steiner, Frankfurt/M.; H. Tadokoro, Osaka; M. Takayanagi, Fukuoka; I. Uematsu, Tokyo; C. Wippler, Strasbourg; H. Zahn, Aachen

Editorial Assistant: A. Heinrich, Springer-Verlag Heidelberg

To cope with the rapid progress of polymer science, a new journal is now published characterized by emphasis on rapid publication of papers containing a most concise description of results.

The character of the new journal is between the purely archival journals of full papers and the so-called "letter journals" consisting exclusively of short communications.

Special features:
- rapid publication of papers
- no page charge
- 50 off-prints of each paper supplied free of charge

Subscription information and sample copy upon request

Send your order to your bookseller or directly to:
Springer-Verlag, Journal Promotion Dept.,
P. O. Box 105280, D-6900 Heidelberg, FRG

North America: Springer-Verlag New York Inc.,
Journal Sales Dept.,
44 Hartz Way, Secaucus, NJ 07094, USA

Springer
International

A. Hebeish, J. T. Guthrie

The Chemistry and Technology of Cellulosic Copolymers

1980. 91 figures, approx. 91 tables. Approx. 500 pages.
(Polymers/Properties and Applications, Volume 4)
ISBN 3-540-10164-0

The driving force behind the great scientific interest in copolymer science and technology is the search for products with useful, new or interesting properties. This monograph provides an informative account of new, improved cellulosic materials and the chemistry and technology involved in their production as well as the first detailed description of grafted and modified celluloses.
The information contained in this book will be of great value to researchers, manufacturers, but also instructors, interested in the modification of cellulosics for textiles, paper, printing, printing inks, paints, and packaging, as well as in polymerization processes and cellulose derivativization.

A. Gandini, H. Cheradame

Cationic Polymerisation

Initiation Processes with Alkenyl Monomers

1980. 12 figures, 9 tables. Approx. 360 pages.
(Advances in Polymer Science, Volume 34/35)
ISBN 3-540-10049-0

This monograph covers the entire spectrum of initiations systems in the cationic polymerisation of alkenyl monomers. Following a detailed outline of the factors which play an important role in determining the behaviour of cationic polymerisation, each type of initiation is discussed individually. Particular emphasis is placed on the two major modes of initiation: initiation by Brønsted acids and initiation by Lewis acids. The authors analyze the present status of this discipline through a critical review of the literature and a series of specific mechanistic proposals, some of which are entirely new. Published material relevant to the understanding of the processes leading to the formation and characterisation of active species is covered exhaustively. The significance of early work is reinterpreted and the impact of more recent studies as well as their shortcomings assessed. The potentials of new experimental techniques are also discussed. Finally, suggestions are offered for future work in many areas on the basis of the mechanistic proposals developed.
This book will help stimulate further ideas, discussions and research in a discipline which is experiencing a lively renaissance.

Springer-Verlag
Berlin
Heidelberg
New York